INSTRUMENTATION FUNDAMENTALS AND APPLICATIONS

INSTRUMENTATION FUNDAMENTALS AND APPLICATIONS

RALPH MORRISON
Instrum
Monrovia, California

A Wiley-Interscience Publication
JOHN WILEY & SONS
New York • Chichester • Brisbane • Toronto • Singapore

Library of Congress Cataloging in Publication Data:

Morrison, Ralph.
Instrumentation fundamentals and applications.
"A Wiley-Interscience publication."
Includes index.
1. Engineering instruments. I. Title.
TA165.M64 1984 681'.2 83-21696
ISBN 0-471-88181-3

Printed in the United States of America

10 9 8 7 6 5 4 3 2 1

PREFACE

"He who proposes to be an author should first be a student."
MONTGOMERY

In all areas of testing and measurement, instrumentation amplifiers play an important role. There are applications such as wind tunnel facilities where hundreds of data channels are needed; in others, a single amplifier may handle a simple medical measurement. The instrumentation amplifier bridges the gap between a transducer or sensor and a recorder or observer. The nature of the signal, the type of data analysis needed, and the signal environment are all variables in the instrumentation problem. This book discusses all these variables in considerable detail.

There is often a wide gap between physical phenomenon and useful measurement, and the instrumentation amplifier may be the only way to bridge this gap: The raw signal has to be amplified and conditioned before it can be correctly recorded and interpreted.

An instrumentation amplifier, however, adds its own errors, and meaningless data can result if it is not correctly used. For example, noise in the form of RFI (Radio Frequency Interference) can completely overload the amplifier. If users do not know this, they may accept nonsense for reality. For self-protection, the user must understand the limitations of the various system connections. This book discusses instrumentation hardware and its proper application so that meaningful measurements can be made.

Many standard transducers generate low-level signals; for example, most strain gages, thermocouples, and quartz accelerometers give only a few millivolts of signal. Before good amplifiers became available, light galvanometers were used to record such transducers directly. In the 1940s, carrier systems were used to amplify strain gage signals. In the 1950s, vacuum-tube chopper-stabilized amplifiers appeared and soon these amplifiers became differential and provided ground isolation. These instruments gave signal gain, but provided little in the way of signal conditioning.

In the 1960s and 1970s instrumentation amplifiers provided filtering, offset, and bridge conditioning. The advent of integrated circuits made these improvements practical. Needless to say, the evolution in instrumentation will continue. With the growing use of computers, instrumentation will change in character to match the demands of lower cost, higher accuracy, and computer interfacing.

A dozen or so small companies in the United States monopolize the instrumentation amplifier business. It takes a small company to handle this kind of activity because there is seldom any large production involved. Each requirement has a different set of specifications, which has forced a profusion of designs and model numbers.

This situation has served to keep the unit price of most instrumentation amplifiers fairly high. Prices in the early 1980s average about $1000 per amplifier channel.

The term *instrumentation amplifier* is not clearly defined in the industry. Manufacturers of integrated and hybrid circuits may label some of their products as instrumentation amplifiers, but the user must add power, connections, gain switching, zeroing, and so on, to operate the device; in effect the user makes his own instrumentation amplifier by purchasing the heart of the design and embedding it in his own hardware. This is a practical approach for a design engineer who cannot use a readymade unit. With added hardware and electronics the device becomes a complete instrumentation amplifer. To get quality results, the user must know the dos and don'ts of instrumentation. This book provides guidance in this area. In a small company, instrumentation design is often handled by one key person, who is often a nondegreed engineer with years of practical experience. I have known four such designers in my career and they were all hard-working creative people who stayed with one company for years. These unique people have done a lot for the military and aerospace sectors by supplying designs and solutions to meet many critical needs. This group of engineers is a very small fraternity and because of its small size, the volume of technical data that has emerged is limited.

I hope this book fills an ongoing need. For more than 30 years in instrumentation, I have spent my time at the workbench as well as in the field. I have designed, built, and delivered hardward to dozens of installations across the country. I have been involved with end users and worked out specifications describing instruments to meet specific needs. This experience has provided the background for the book.

My first book, *Grounding and Shielding in Instrumentation,* published by John Wiley and Sons in 1967 (second edition, 1977), looks at instrumentation from the shielding point of view. This new book continues the story and delves more deeply into the inner workings of the transducer, the amplifiers, and the instrumentation environment. As the world gets more complex and more emphasis is placed on performance, reliability, and cost, the task of the instrumentation engineer will definitely increase in scope. It is my hope that this book will contribute to making the task easier.

The opening three chapters deal with transducers and transducer connections. If a data channel is to operate effectively, all aspects of the signal process must be properly considered. Input considerations receive extensive treatment here because they are usually the most critical.

Chapters 4 and 5 deal with the input circuitry at the amplifier. Such topics as RFI, transducer excitation, multiplexing, and calibration are discussed. Chapters 6 and 7 deal with amplifier performance and include specification discussions, common-mode types, signal interference, and power isolation.

The last chapters discuss digital control, testing calibration, and the future in instrumentation.

The book is written to be general purpose. The instrumentation discussed can be located aboard aircraft or be used in ground testing of aircraft structures. It can be used for medical research, studies of structural dynamics, or automobile safety. It can be used underwater or in space. The range of instrumentation application is endless, but the problems and processes have many common denominators. Instrumentation is presented in general terms so that all users may benefit.

I wish to thank Roland C. Hawes for suggesting that this book be written. Many thanks to Barbara Bott for her very fine job of typing, and a thank you to my wife Lee, who again put up with the trials of an author.

RALPH MORRISON

Pasadena, California
January 1984

CONTENTS

CHAPTER ONE

BASICS

"What is the use of running when we are not on the right road?"

It is necessary to treat a few basic concepts before discussing general instrumentation. Words such as *ground* and *earth* need to be defined. The elements of electrostatic shielding need to be included so that later discussions can be effectively presented. Because transformers and transformer shielding are often misunderstood, this area needs treatment. It is necessary to place words such as common mode and differential mode in context. A few words about radio frequency (rf) are also relevant.

Readers tend sometimes to skip to topics of interest rather than start at the beginning. However, it will help the reader to read carefully this first chapter before proceeding to other areas of interest.

1.1 THE ELECTRIC FIELD

All potential differences result in an electric field. The field starts on conductors of the most positive potential and terminates on points of lower potential. If a conductor has no IR voltage drop along its length, it will have no tangential component of the electric field. Viewing circuitry in a field sense, conductors are used to place a proper electric field onto each electric component.

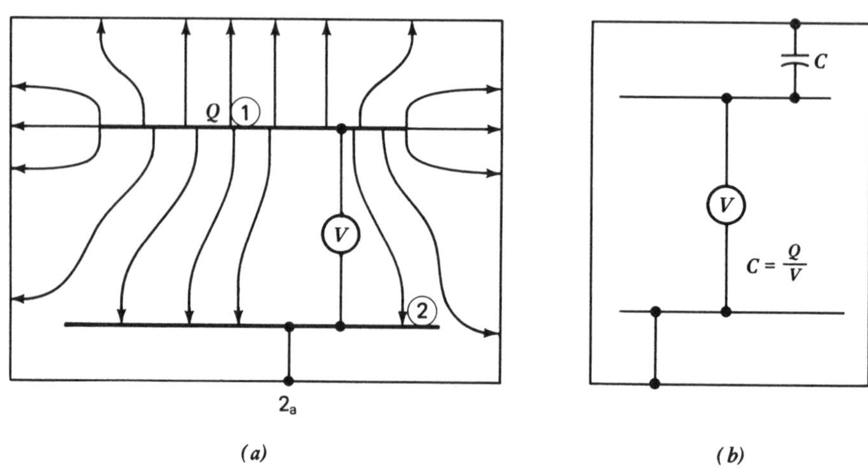

FIGURE 1.1 *(a)* The *E* field and *(b)* its equivalent capacitance.

If a circuit is completely enclosed by a metal box (shielded), the electric field *E* external to the box will terminate on the outside of the box. This field does not appear inside* the box and does not influence circuits within the box. In concept this box can be any shape and can extend over signal conductors and power sources. Unfortunately, this perfect world is not available because conductors must penetrate the shield to make circuits useful.

It is impractical to draw *E* fields to illustrate how a typical circuit functions. Thus the conventional method involves the use of component symbols to show circuit geometry. When parasitics play a role, symbolic capacitors are used to illustrate the presence of fields. All *E* fields *start* and *terminate* on charges. When the potentials in the circuit change, the *E* fields change and the charges on the various conductors must readjust to new values. This motion of charges is simply current flow. A capacitor also equates a change in potential to a current flow and thus the symbology is technically correct. Figure 1.1 shows the *E* field and symbolic capacitors for a simple circuit.

Note that in Figure 1.1*a* very little of the *E* field terminates on the bottom surface **2a.** This would imply that very little current flows to the bottom surface as *V* changes. The symbology in Figure 1.1*b* shows current flow but does not show how it is distributed. Note that if a charge *Q* resides on a conductor **1** that a charge -*Q* resides on conductor **2.**

*This assumes that surface currents will flow on the outside of the box.

1.2 SELF- AND MUTUAL CAPACITANCE

The single capacitor shown in Figure 1.1*b* does not represent the real world because multiple conductors are the rule in circuits. Viewed in the field sense, portions of the field terminate on conductors where current flow is irrelevant. Fields that terminate on sensitive conductors need to be clearly identified.

In the physics of electrostatics, two kinds of capacitances are discussed. A self-capacitance is defined as the ratio of voltage to charge on one conductor, all other conductors being at zero potential. The ratio of potential on one conductor to the charge induced on a second conductor with that conductor and all others at zero potential is called mutual capacitance. It is this mutual capacitance that often interests the engineer. Figure 1.2 shows an E field and a capacitance representation of self- and mutual capacitance. The self-capacitance of conductor **1** is labeled C_{11}. The mutual capacitance between conductor **1** and conductor **2** is labeled C_{12}. Note that the use of double subscripts C_{11}, C_{22}, and C_{33} designate self-capacitances; C_{12}, C_{31}, C_{23}, and so on, designate mutual capacitances. It is interesting to note that mutual capacitances are always negative and that $C_{12} = C_{21}$, $C_{32} = C_{23}$, and so on.

The structure shown in Figure 1.2 could easily represent a shielded cable with conductor **3** the shield. If conductor **2** is at a potential that could contaminate the signal on conductor **1**, then the value of capacitance C_{12} allows a simple measure of the pickup. The equivalent circuit is shown in Figure 1.3. The impedance Z represents the source and/or terminating impedance on the signal line. The shield **3** in Figure 1.2*a* is in effect guarding against the potential V. C_{12} is also called a leakage capacitance.

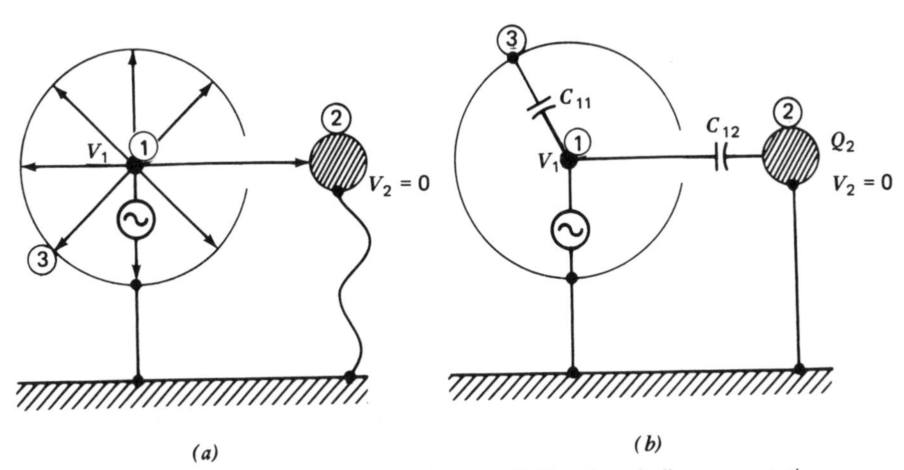

FIGURE 1.2 Self- and mutual capacitances—field and symbolic representation.

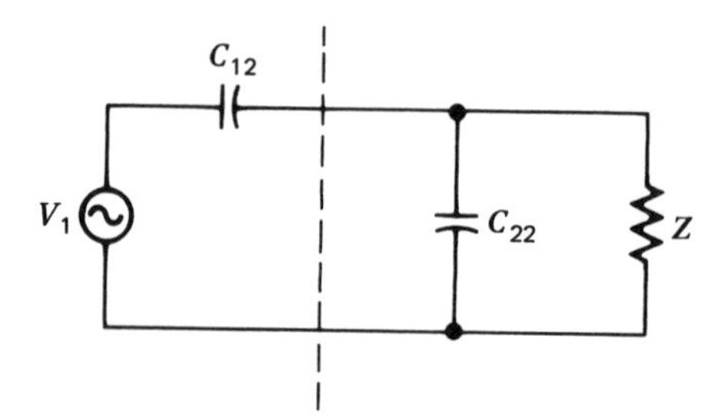

FIGURE 1.3 A circuit equivalent of a mutual capacitance.

1.3 THE SHIELD POTENTIAL PROBLEM

Although a circuit fully enclosed by a shielded enclosure is protected against external field influences, it may not function properly. Figure 1.4*a* shows how the shield can act to couple signals in an unwanted manner. The equivalent circuit is shown in Figure 1.4*b*. This class of coupling or feedback can cause circuit instability. To avoid this coupling, capacitance C_3 is usually shorted out by tying the zero-signal reference conductor to the shield. Figure 1.5 shows this circuit. This is commonly known as "ground the shield."

Consider the undefined shield enclosure again. If one conductor is brought out of the shield enclosure and tied to an external structure (ground), parasitic capacitances can couple unwanted signals directly into the circuit, as shown in Figure 1.6. Capacitances C_1 and C_2 in series couple an unwanted signal V directly into the amplifier through capacitances C_3 and C_4. Again, if capacitance C_2 is shorted out, this coupling path is removed.

There are several choices for the location of this shield-to-reference conductor connection. To illustrate the correct connection consider the circuit

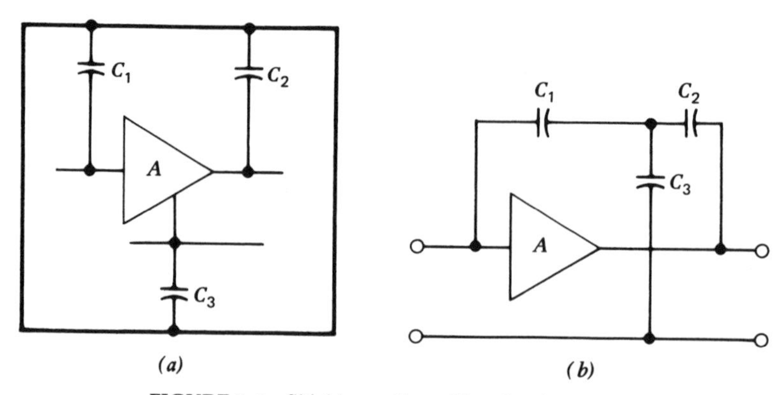

FIGURE 1.4 Shield coupling with a floating shield.

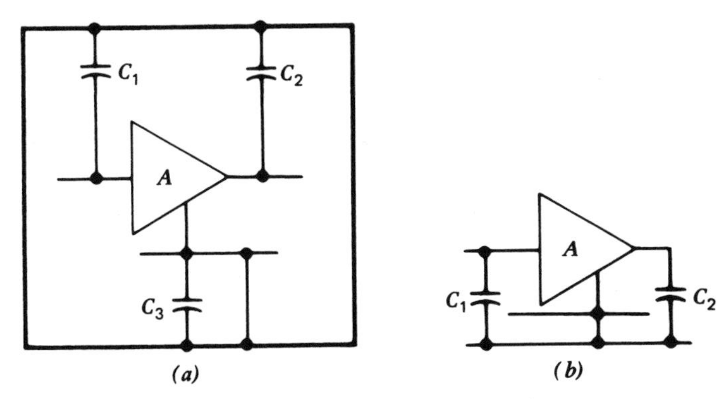

FIGURE 1.5 A nonfloating shield.

in Figure 1.7. Here conductor **5** might be the low side or the return side of a low-level signal that is to be amplified. Any external potential V_1 on conductor **2** with a capacitance to the shield can cause current to flow in the loop **1–2–3–4–5–6–1.** Current flowing in conductor **5** can be troublesome if it adds signal to the amplifier input. To avoid this problem **3** must be directly connected to **5** as shown in Figure 1.8. The unwanted current now takes the shield path **1–2–3–7–6–1** and does not flow in conductor **5**. Note that 100 μA of current flowing in 1 ohm of lead wire will cause a noise pickup at 100 μV. These two rules emerge from this discussion:

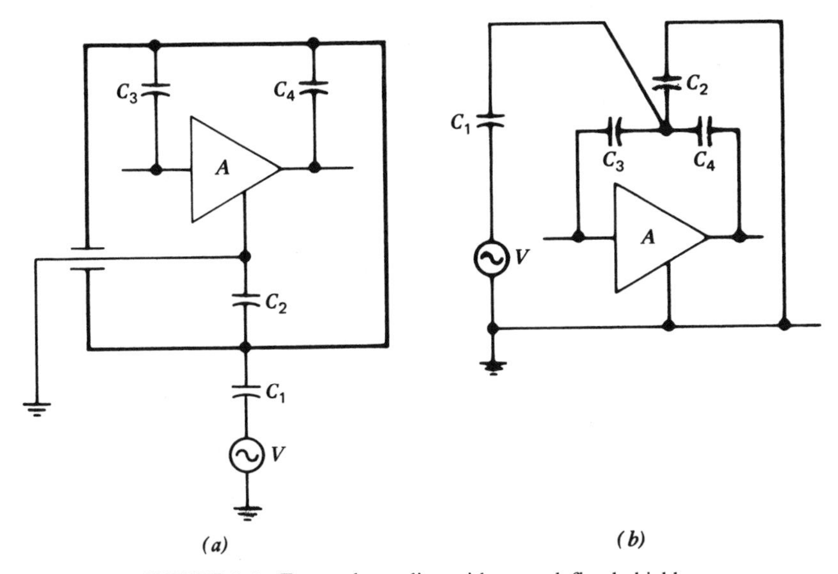

FIGURE 1.6 External coupling with an undefined shield.

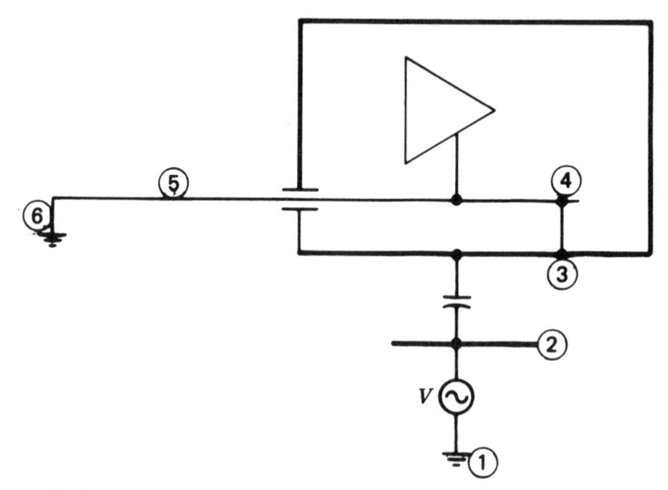

FIGURE 1.7 A single conductor exiting a shielded circuit.

1. Shields must be connected to a zero-signal reference potential to avoid unwanted coupling.
2. This connection must be made so that parasitic currents do not flow in signal conductors.

Both rules can be followed by connecting the signal shield to the zero-signal reference conductor at the point of signal origin.

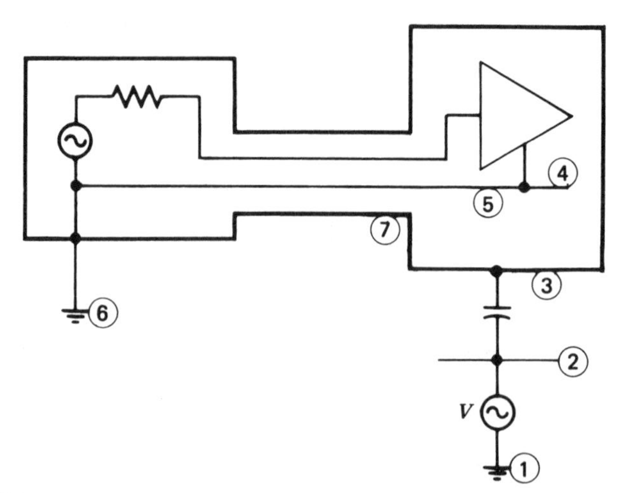

FIGURE 1.8 A correct shield to reference conductor connection.

1.4 GROUNDS

The words *ground* and *earth* are often used to describe electrical connections to structures, frameworks, racks, and so on. Earth usually describes an electrical connection to the soil under a structure for lightning and safety protection.

Signals are usually sensed on metal structures that are tied to earth for safety. Obviously, the skin of an aircraft or missile in flight is not earthed. For instrumentation purposes, any conducting surface associated with a signal source or a signal termination is a ground. Conduit, raceways, neutrals, racks, skins, and points along the earth's plane are all considered grounds.

No two ground points are at the same potential. Currents from power line leakages, load unbalances, various rf transmitters, and so on, all contribute to ground potential differences. It is not practical to short ground points together to force a zero-potential difference because these connections are inductive and do little good at ac. Instrumentation must operate on the assumption that there will be ground potential differences between all signal connecting points.

Grounding philosophies have their origin in electrical safety. If a motor winding shorts to an ungrounded cage, it represents a definite user hazard. Neutrals must be earthed at building entrances to protect against lightning. The multiple grounding of shields for safety purposes follows in this general path. Because of ground potential differences a signal shield *cannot* be multiply connected to a ground if it is to function as an electrostatic guard against foreign potentials. Two or more shield ties will force a voltage gradient along the shield. A proper shield must be at one potential along its entire length.

1.4.1 Safety

Safety rules must always be observed. Grounds cannot be lifted to accommodate instrumentation processes. Rather the instrumentation must accommodate the world of real ground potential differences. Signal sources should be connected to their ground, and signal outputs must be grounded at recorders or computers. The signal leads that connect to a ground should not be labeled "earth" or "ground" because building codes may force a safety treatment. Labels such as signal return, signal common, and guard can be used to avoid this problem.

1.4.2 Earth Connections "Clean Grounds"

Low-resistance earth connections are difficult to obtain. For steel rods driven into salty wet earth and a large surface area grid, the lowest resistances are on the order of a few ohms. Earths of this quality are desirable for safety reasons but rarely achievable. There is a school of thought that requires a low-resistance earth connection for a system ground to reduce system noise. Unfortunately, this approach rarely serves this purpose. The impedance from the system ground to "earth" is quite inductive. The first user to drain unwanted current to earth contaminates the connection; the second user does not have a "clean ground."

The presence of an E field from any rf transmitter causes currents to circulate in *all* loops. The earth connection discussed above is often a segment that shares many such loops. The resulting current flow causes an IR voltage drop in this earth-connecting segment and the clean ground is contaminated before it is used. The moral to this story is simply this: The clean ground approach to reducing noise is a myth.

1.5 LIGHTNING

Lightning strikes can cause ground currents that exceed 50,000 A. As a result, earth gradients in excess of 1000 V/m can result. These gradients can cause arcing and severe damage to electrical equipment unless circuits are protected. In a properly protected building, lightning currents are drained to earth on a separate set of drain conductors. If these currents use the building steel to get to earth danger to personnel and equipment within the building is possible. Even when separate conductors are used, fields around the lightning drain lines can still couple large currents into the building steel. These induced currents can be large enough to cause damage to equipment within the building unless precautions are taken.

1.6 THE GENERAL INSTRUMENT PROBLEM

The rules developed in Section 1.3 require a shield-to-zero-signal reference connection at the signal ground point. The usual problem is to sense a signal associated with one ground and terminate a conditioned version of this signal at a second ground. Because these grounds are assumed to have a potential difference, the conditioner or instrument must reject this potential difference.

Figure 1.9 shows the general instrumentation problem. The ground potential difference must not allow current flow in signal lines or the potential difference will be converted to normal-mode or differential-mode signal. The voltage E is usually referred to as common-mode signal because it is impressed in common on both input leads.

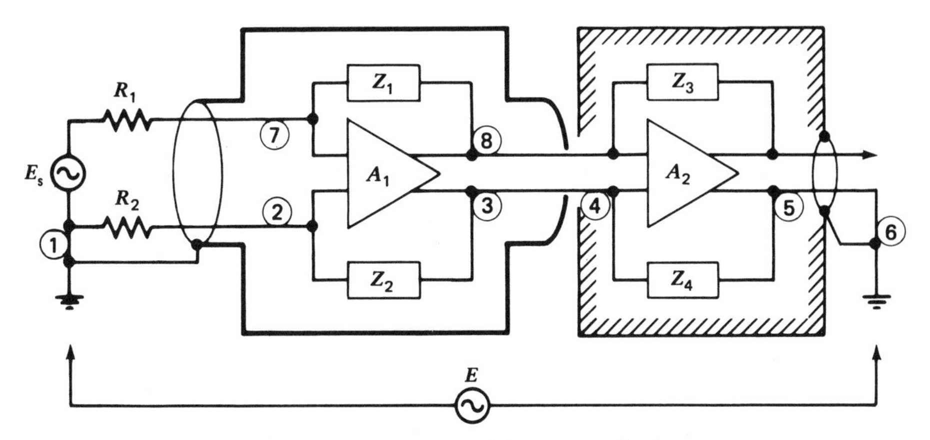

FIGURE 1.9 The general instrument circuit.

The two gain blocks A_1 and A_2 in Figure 1.9 are powered by supplies referenced to grounds **1** and **6**, respectively. The coupling problem being considered involves the input impedances Z_1, Z_2 or Z_3, Z_4. Assume impedances Z_3, Z_4 are much greater than Z_1, Z_2. Then voltage E will cause current to flow in loops **1–2–3–4–5–6–1** and **1–7–3–5–6–1**. The current is limited by the value of Z_3 and Z_4.

Assume E is 10 V, $A_1A_2 = 1000$, $R_1 - R_2 = 1$ kohm, and the maximum error signal at the output is 10 mV. The current flowing in R_1 and R_2 is limited to 10 nA. This forces $Z_3Z_4/(Z_3 + Z_4)$ to be greater than 10^9 ohms. At 60 Hz this is only 3 pF. The shield around amplifier A_1 would shunt Z_1 and Z_2 with more than 3 pF of capacitance. These parasitic capacitances would thus negate Z_1 and Z_2 if they were above 10^9 ohms.

Instrumentation between grounds requires one high-impedance link or excess errors can result. Because the two-amplifier design shown in Figure 1.9 is not economical, the one-amplifier approach in Figure 1.10 is frequently used. Note that the input shield is connected only at the source and is not connected within the amplifier. The shield must enter the amplifier to protect leads **1** and **2** so that mutual capacitance C_{13} and C_{23} are both held below 1 or 2 pF. This means that great care must be taken to protect leads **1** and **2** through every bulkhead or connector.

The instrument shown in Figure 1.10 is usually referred to as a Differential Instrumentation Amplifier. The differential signal E_s is amplified and the common-mode signal E is rejected. The differential inputs to an oscilloscope can also reject common-mode signals, but the input impedance using a 1:1 probe is only 10^6 ohms shunted by 30 pF. Thus an oscilloscope does not meet the requirements of an instrument amplifier as defined above.

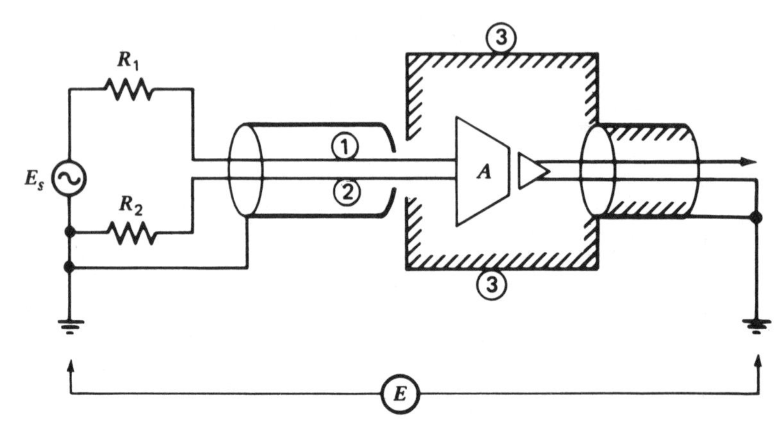

FIGURE 1.10 A single amplifier to reject signal E. The split triangle implies a circuit with more than one IC.

1.7 COMMON MODE AND DIFFERENTIAL MODE

The circuit in Figure 1.10 is designed to reject common-mode signals defined as voltage E. This voltage is a common-mode signal because it is impressed in "common" on both input leads. The signal E_s is a differential-mode signal, for the amplifier responds to this "difference" signal. Frequently, this signal is called a normal-mode signal. In the telephone industry longitudinal and common-mode are equivalent terms and transverse and differential mode are equivalent terms.

A second type of common-mode signal frequently encountered in instrumentation is the excitation voltage used in a strain gage bridge. If one corner of the bridge is grounded, then one-half of the excitation is common-mode and must be rejected. This circuitry is shown in Figure 1.11. Noise can enter the amplifier in Figure 1.11 in many ways. If rf current flows along the shield, it can be converted to both differential- and common-mode input signal. The treatment of these unwanted signals is discussed later.

A common-mode signal for an instrument amplifier is defined as the average input signal or

$$E_{CM} = \tfrac{1}{2}(E_1 + E_2)$$

and the differential-mode signal is defined as the difference voltage

$$E_{DM} = E_2 - E_1$$

Note that if $E_1 = 0$, the difference signal is E_2 and the common-mode signal is $1/2\ E_2$.

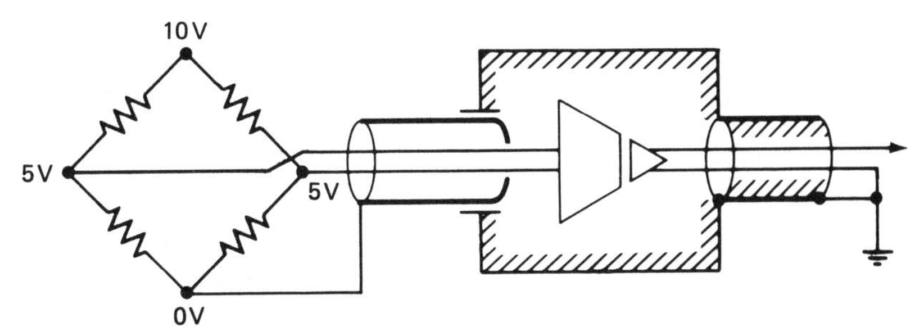

FIGURE 1.11 Excitation voltage as a common-mode signal.

1.7.1 Power Common Mode

Power line voltages can also be considered in the common-mode and differential-mode sense. The normal power voltage is a differential-mode signal. If both sides of a single-phase line move together with respect to a system ground, this is a power line common-mode signal. This type of signal is apt to be transient in nature and can be disruptive to low-level instrumentation or logic systems.

1.8 TRANSFORMERS

The problems created by the need to power equipment are numerous. Transformers and their power-related signals are often the bane of the designer or user. The problems result because the grounds associated with power sources are brought up against or brought inside the shields used to protect low-level signals.

Electrostatic shields can be placed between coils in a transformer to reduce the problem. The shield is shown symbolically in Figure 1.12. Iron core transformers are usually drawn schematically with solid lines between the coils, but have been omitted for clarity.

In practice, the coils in a transformer are placed one on top of the other to minimize the leakage inductance. The shield shown in Figure 1.12 is usually a thin layer of metal between two coils. A cutaway view of such an arrangement is shown in Figure 1.13. Note that the shield is insulated at the overlap to avoid a shorted turn.

This type of shielding can still leave a mutual capacitance C_{12} between primary and secondary coils of about 10 or 20 pF. If the shield is folded on the edges of the secondary coil and if the primary leads are dressed away

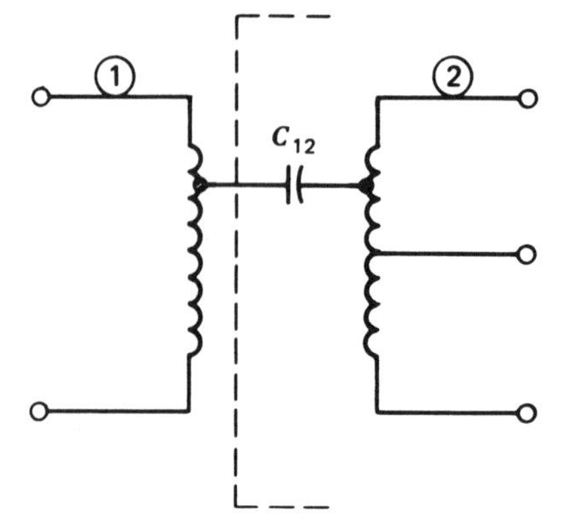

FIGURE 1.12 A shield between coils in a transformer.

from the secondary leads, this mutual capacitance can be reduced to below 1 pF. If the secondary leads are shielded as they exit the coil, the mutual capacitance can be held to a few tenths of a picofarad.

1.8.1 A Floating Supply

A floating power supply can be used to illustrate the single transformer shield problem. In Figure 1.14 the shield is connected to the reference conductor of the power supply. The ground at point **1** is not at the same potential as ground **7**. The primary coil has the start or the finish near the shield so that a fraction of the primary voltage appears in series with capacitance C_{23}. The resulting current loop is **1–2–3–6–7–1** and this current flows in R_2. This IR drop is undesirable in a floating power supply. If the shield is connected to **1** or to **5** a similar problem exists. A single shield is inadequate to control the currents generated by coil voltages.

When a second shield is introduced, the problem appears more tenable. See Figure 1.15. The primary voltage V_{12} circulates current in loop **1–2–3–**

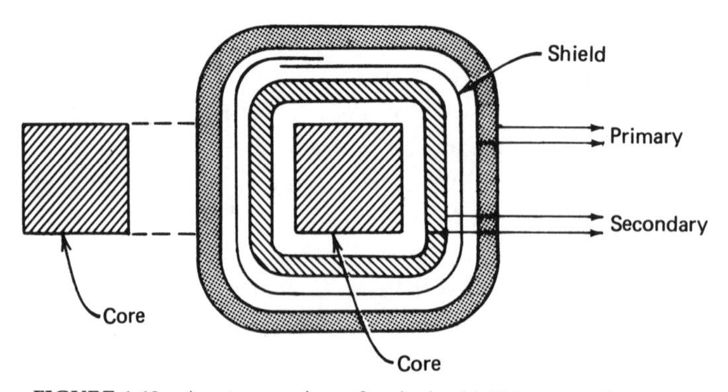

FIGURE 1.13 A cutaway view of a single shield in a transformer.

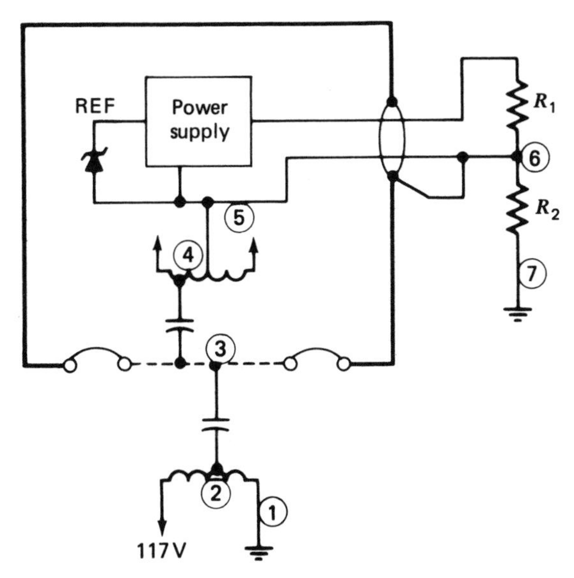

FIGURE 1.14 A floating power supply shield problem.

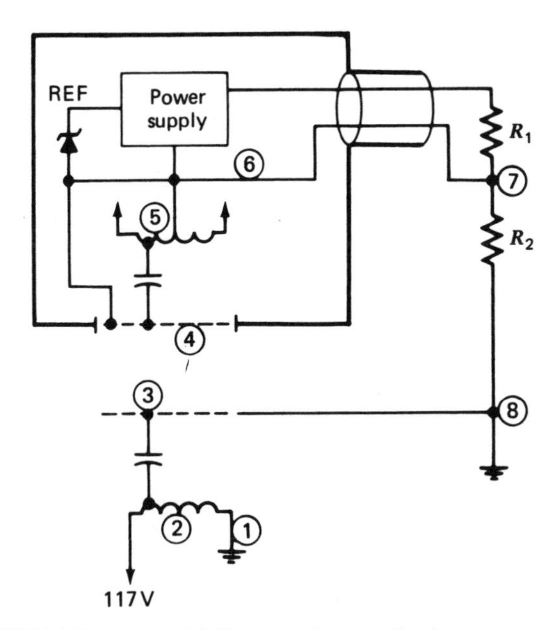

FIGURE 1.15 A two-shield approach to the floating power supply.

8–1 and this current does not flow through resistor R_2. Secondary voltages circulate current in loop **4–5–6–4**. Of course, if mutual capacitance C_{24} is not kept low, primary voltages can circulate current in loop **1–2–4–6–7–8–1** and this current flows through resistor R_2. This implies that the primary shield should be a tight or so-called box shield.

Transformer shields are thus useful in controlling the flow of unwanted power currents. When these currents flow in cable shields or between external grounds and not in signal conductors, the transformer will not cause trouble.

Leakage capacitances such as mutual capacitance C_{24} can be measured by using a well-shielded oscillator to drive the primary coil. All other coils are grounded and a small resistor senses the current to the secondary shield. This circuit is shown in Figure 1.16. If E_s = 10 V at 10 kHz, a 1 mV signal across R represents a capacitance C_{24} of 0.15 pF. This measurement may not compare with the current sensed in R_2 in Figure 1.15 when the circuit is operational because

1. A small amount of transformer flux can couple into the shield-coil loop. This flux capture causes a series emf that adds to the loop current. This flux may appear during portions of the cycle when the transformer is heavily loaded.
2. A shield is a single turn. The resulting emf can appear in series with the shield and cause loop current to flow.

If mutual capacitances as low as 0.15 pF are to be maintained, lead dress and shielding within and outside the transformer are critical. A simple transformer header is usually unable to protect this level of capacitance.

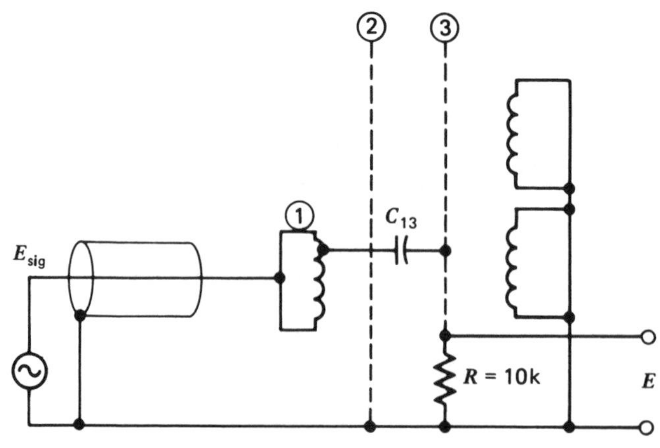

FIGURE 1.16 Measuring a leakage capacitance.

1.9 THE RF PROBLEM—INTRODUCTION

The low-frequency problems in instrumentation involve the *E* field. When this field changes, currents must flow and these currents create a magnetic or *H* field. For many geometries the presence of both an *E* field and an *H* field can result in electromagnetic radiation. At low frequencies only a small percentage of the stored field energy can be radiated. As the frequency increases and as the dimensions of the current paths approach a fraction of a wavelength, the radiation processes become more efficient. Radiation and rf suseptibility show reciprocity. If the fields are known in the vicinity of a conductor, the currents in that conductor are uniquely defined. Thus a field measurement can provide data to indicate the voltages and currents that can flow in various cable geometries. Note that a transmitting antenna can easily substitute for a receiving antenna.

At frequencies above 10 MHz even small capacitances appear as low reactances. For example, 100 pF is 150 ohms at 10 MHz. This means that nearly every conductor associated with a rack of instrumentation forms a ground loop with the ground plane and numerous other conductors. No pair of conductors is immune. Grounds, neutrals, commons, shields, power lines, control lines, and signal lines all form loops. Radio-frequency currents in these loops can be very troublesome.

Ground loops rarely match the impedance of free space. For this reason field levels are altered by the presence of a loop. As a first approximation, ambient field strengths can be used to calculate possible impact on systems performance.

Many radiators are in our midst. Typical rf sources are radio, television, and radar sites. Also present are such devices as diathermy machines, arc welding, fluorescent lights, and glow lamps. These latter devices radiate a wide spectrum of noise. Another class of radiated energy stems from transient phenomena caused by contact arcing, surge or inrush current, step loads, and so on. The intermittent nature of this phenomenon makes it difficult to isolate if it is causing trouble.

Once rf energy is captured in any transmission line or loop, it is called conducted rf energy. This resulting current still has an *E* field and an *H* field that can continue to interact with other conductors in the system.

1.9.1 The Far Electric Field

For distances greater than $\lambda/2\pi$ where λ is the wavelength in free space, the *E* field in volts per meter from a short segment of wire or doublet becomes

$$E = \frac{Z_0 ID\pi\lambda}{2r}$$

where Z_o = impedance of free space = 377 ohms
λ = the wavelength in meters
D = length of wire in meters
r = distance from wire in meters
I = current in amperes

To place this in perspective the following field strengths are typical at 1 km from a radiating source.

Paging system	0.05 V/m
AM broadcast	1 V/m
FM broadcast	3 V/m
UHF television	11 V/m
Weather radar	200 V/m

1.9.2 The Near Electric Field

The E field at distances less than $\lambda/2\pi$ from a segment of wire or doublet is given roughly by the equation

$$E = \frac{Z_0 ID\lambda}{8\pi^2 r^3}$$

This field might be present on a printed circuit board or even within an integrated circuit.

1.9.3 The Near Magnetic Field

Magnetic fields near a conductor are classed as near fields for distances r less than $\lambda/2\pi$. The near field is given approximately by the expression

$$H = \frac{ID}{4\pi r^2}$$

where I = current in amperes
D = loop diameter in meters
r = distance from loop in meters.

This field might be present on a printed circuit board or even within an integrated circuit.

1.9.4 Loop Coupling

The coupling to any loop is dependent on loop area. If the ground plane is involved, it can be the reenforcing bars in a concrete slab, a metalized floor, or simply the earth. When the earth is involved, the average depth of penetration of the field must be considered. This varies with frequency and soil conditions.

As a first approximation the tangential E field is zero near a ground plane. The voltage V is an open circuit formed by a cable and the ground can be found by considering the magnetic flux intercepted by the loop area. The E field and the H field are related to the impedance of free space, or E/H = 377 ohms. The component of H that is of interest is along the length of the cable run. The voltage will be roughly proportional to the cable length times the height of the cable over the ground plane. The resulting open circuit voltage will be

$$V = \frac{2\pi Hhlc}{\lambda}$$

where V = voltage
H = magnetic field in webers (10^4 gauss = 1 weber)
h = height above ground in meters
l = cable run in meters
c = velocity of light (3×10^8 m/s)
λ = wavelength in meters

This voltage can be used to calculate the current that flows in the cable/ground-plane loop.

CHAPTER TWO

TRANSDUCERS

"All bread is not baked in one oven."

A measurement always depends on some devices to convert a parameter to an observable phenomenon. Devices that perform this conversion are called transducers. In this book transducers are limited to those types that convert a parameter to an electrical signal.

Many measurements are by their nature difficult to make. The choice of transducer, the way it is applied, and the methods of calibration all are key to getting good results. The problem, however, goes much deeper, for users must mount the transducer and wire it to their instrumentation. They must also select cabling, shielding, ground points, and instrumentation hardware. These considerations are as important as the transducer selection itself.

Measurements are truly systems problems. No one element in the measurement chain can stand alone. Because signals start at the transducer it makes sense to treat this area early. This discussion is aimed at some of the interface problems that might limit system performance. This chapter discusses signal levels, source impedances, bonding, and parasitic capacitances for each transducer type. A full treatment of transducer performance is outside the scope of this book.

2.1 STRAIN GAGE DEVICES

A solid under stress will undergo a mechanical deformation. This dimensional change is called strain. This strain can be measured optically, mechanically, or magnetically, but most frequently by using a strain gage. A strain gage is a resistive element that is cemented to the solid under test. When the solid is strained, the resistive element undergoes a similar strain. The resulting change in resistance is then a measure of strain on the structure being tested. This type of element is called a bonded strain gage.

The change in resistance in a strain gage is usually small. To measure this change, the resistive element is usually connected as one arm of a Wheatstone Bridge. When the gage is in a neutral strain condition, the bridge is balanced. Any contraction or elongation of the gage causes the bridge to "unbalance" and an electrical signal results. A typical Wheatstone Bridge is shown in Figure 2.1. The bridge is excited by a voltage E, which is typically about 10 V; E_{sig} may be as large as 15 mV. It is given by the expression

$$E_{sig} = \frac{E\Delta R}{4R} \tag{2.1}$$

Strain gages are manufactured using several technologies. A flat wire grid cemented to a base is perhaps the most common approach. The base is in turn cemented to the structure under test. A second method is a metal foil. Here the foil is etched to form the resistive grid that makes up the gage. In some applications a grid of metal is vacuum deposited on a substrate to form the gage element. Finally, gage elements can be made of specially doped semiconductor materials. These gages can produce large signals, but they can be very temperature sensitive.

Strain gage designers try to optimize resistance changes per unit strain. This usually requires grid wire diameters from 0.02 to 0.06 mm or metal film

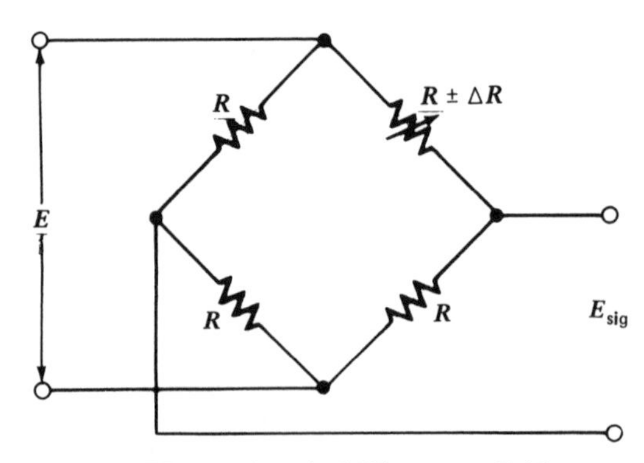

FIGURE 2.1 A typical Wheatstone Bridge.

thicknesses of a few microns. The temperature coefficient of the element is critical. If the coefficient is too high, then resistance changes due to temperature cannot be differentiated from changes due to strain. The problem also includes the bond between the gage element and the connecting leads because this junction is a thermocouple. Gages are available to handle different testing environments and their sensitivities can vary over a wide range.

The sensitivity of a gage element is measured as a ratio of percentage resistance change to a percentage strain. The gage factor GF is given by the expression

$$GF = \frac{\Delta R/R}{\Delta L/L} \tag{2.2}$$

where $\frac{\Delta R}{R}$ = percentage resistance change

$\frac{\Delta L}{L}$ = percentage strain

Gage factors for bonded strain gages vary from 0.5 to 6. Typical gage factors are around 2 or 3. Semiconductor gages can have gage factors from 50 to 200, but their application is limited by temperature considerations. The full-scale strain that can be accommodated by a gage varies. A typical full-scale strain might be 0.001 in./in. If the gage factor is 6, the ratio of $\Delta R/R$ is 0.006. By using equation 2.2, the resulting full-scale bridge signal would be 15 mV. The grid pattern of a simple gage element is shown in Figure 2.2. The gage

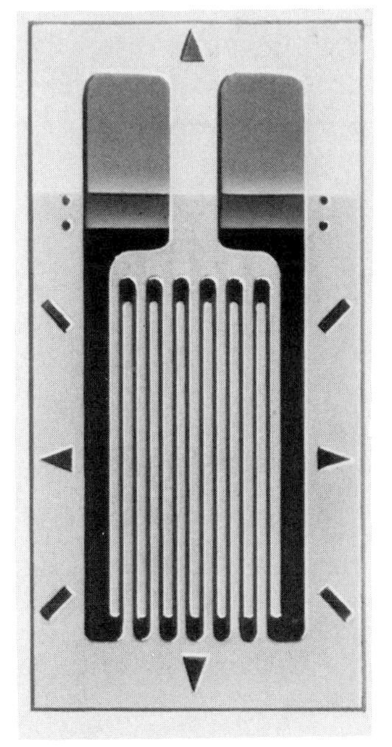

FIGURE 2.2 A grid-type strain gage. (Photograph courtesy of Micro-Measurements Division, Measurements Group, Inc., Raleigh, North Carolina.)

resistance changes when there is a resistance change in the Y direction only. Thus the gage can be cemented to a structure to sense a particular component of strain. The strain measured by a gage is a strain averaged over the length of the gage. For this reason gages are available in various geometries to accommodate various applications.

2.1.1 Multiple Active Arms

It is common practice to use two or even four gages in the Wheatstone Bridge configuration as shown in Figure 2.3. The arms of the bridge that contain strain gages are called active arms. When two gages are used, the resistance changes can be in the same or in opposite directions. These relative changes can be used to enhance or cancel signals depending on the application. If the change in resistance ΔR is in the same direction for two gages, these gages can be placed either in opposite arms to enhance the signal or in adjacent arms to cancel the signal. If the changes in resistance ΔR are in the opposite direction, the gage locations are reversed to enhance or cancel the

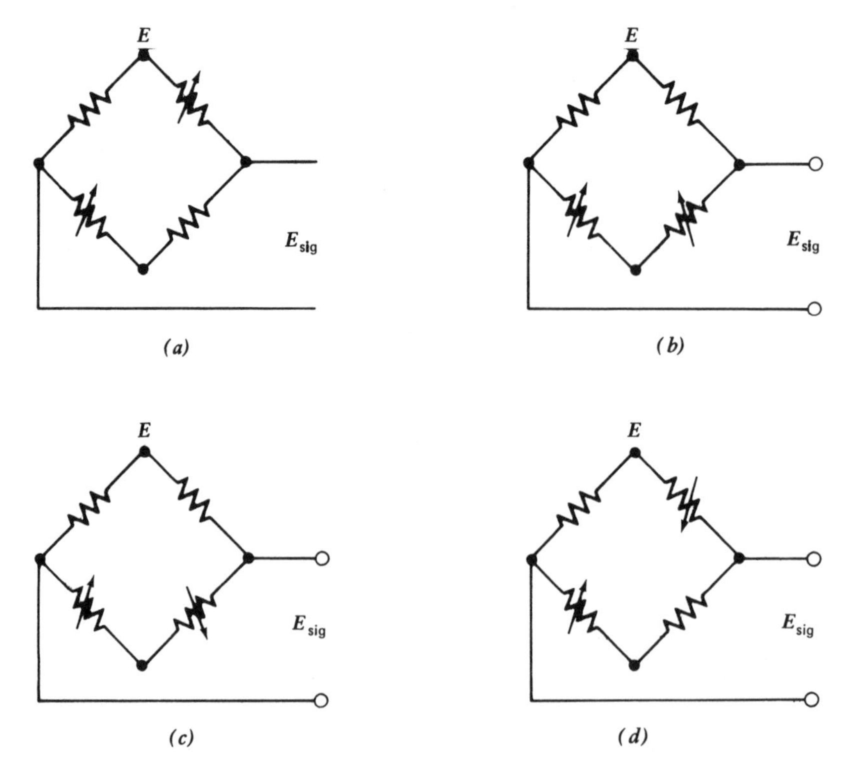

FIGURE 2.3 Various two-active arm bridge configurations: *(a)* signal enhancement; *(b)* signal cancellation; *(c)* signal enhancement; *(d)* signal cancellation.

signal. See Figure 2.3 for the various gage configurations. The bridge signal is given by equation 2.3:

$$E_{sig} = \frac{\Delta RE}{2R} \tag{2.3}$$

Figure 2.4 shows how this active arm could be used to separate elongation from bending in a long bar. The gages are placed on opposite sides of the bar. When the bar is bent, the two gages produce resistance changes in opposite directions. Any component of strain that elongates both gages will produce a signal in the bridge. The active gages would be connected as in Figure 2.3a. Obviously, gages can be located on a structure to separate components of strain or separate shear from strain.

When all four gage elements contribute equally to a particular signal, the bridge signal is given by

$$E_{sig} = \frac{E\Delta R}{R} \tag{2.4}$$

2.1.2 Gage Mounting

Strain gage mounting techniques are not discussed here. It is important, however, to note that a gage must be in intimate contact with its mounting surface if it is to elongate with the structure. This implies a distributed capacitance from the gage to the structure that can be as much as 100 pF. This capacitance can be a source of noise at high frequencies. If only one bridge arm is active, the noise signal is sensed unsymmetrically. If two adjacent bridge arms are active, there can be a first-order noise canceling effect. This noise results from a potential difference between the structure being measured and other grounds in the system. If this potential difference carries current flow in one gage resistance through its mounting capacitance, the resulting IR drop is converted to input signal. A 10 V signal at 10 kHz causes about 30 uA to flow in 50 pF (this is 1.5 mV in 50 ohms). The resulting input signal is about 0.3 mV or 2% of a typical full-scale signal. Single active arm bridges are prone to noise pickup unless the structure is used as the input ground (see Section 3.1). A typical gage installation is shown in Figure 2.4a.

Strain gages have resistances from 50 to 1000 ohms. Excitation levels are limited to about ¼ watt per arm to avoid overheating. Obviously, the greater the excitation level the larger the resulting signal. In cases in which signal levels are inadequate, a pulsed excitation system can be used. One very effective method for short tests is to operate the bridge at a lower excitation level before and after the test and raise the excitation level during the test. This technique can provide signal level enhancement without undue overheating for short tests.

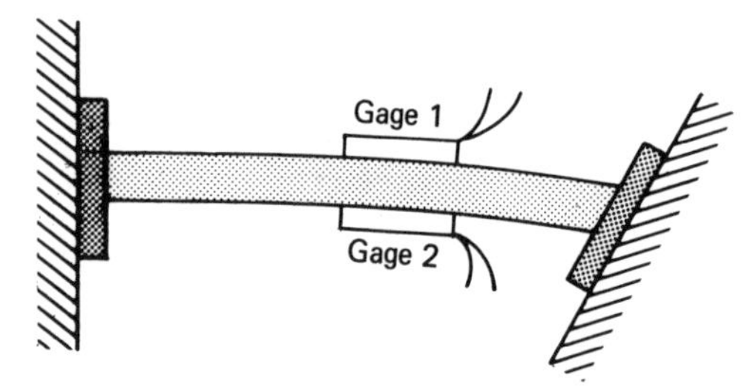

FIGURE 2.4 Two gages applied to a bending bar.

2.1.3 Bridge Excitation

A strain gage bridge can be excited from a constant voltage source or from a constant current source. The excitation source can be either an integral part of an instrument amplifier or a separate voltage source that excites one or more gages. The problems associated with multiple gages, shields, and grounds are covered in the next chapter. The problem gets even more complex when remote excitation sensing and calibration processes are considered.

FIGURE 2.4*a* A typical gage installation. (Photograph courtesy of Micro-Measurements Division, Measurement Group, Inc., Raleigh, North Carolina.

TABLE 2.1
Output Signal per Unit of Bridge Excitation Plus Second-Order Error Terms for Various Bridge Configurations[a]

	One Active Arm	Two Active Adjacent	Two Active Opposite	Four Active Arms
Constant voltage	$\frac{\Delta R}{4R} - \frac{\overline{\Delta R^2}}{8R}$	$\frac{\Delta R}{2R} + \frac{\overline{\Delta R^3}}{8R^3}$	$\frac{\Delta R}{2R} - \frac{\overline{\Delta R^2}}{4R^2}$	$\frac{\Delta R}{R}$
Constant current	$\frac{\Delta R}{4R} - \frac{\overline{\Delta R^2}}{16R}$	$\frac{\Delta R}{2R}$	$\frac{\Delta R}{2R}$	$\frac{\Delta R}{R}$

[a]Note the number of cases where the signal is linear with strain using constant current excitation.

Constant current excitation has two advantages. First, the excitation level at the gage element is independent of lead length and second, most bridge configurations provide a more linear output signal versus strain. A tabulation of the second-order error terms for various bridge configurations is given in Table 2.1.

The linearity provided by constant current excitation is lost at high frequencies for long cables. This occurs because the gage discerns the current source in parallel with the cable capacitance. The result is an excitation that crosses over from constant current to constant voltage at some midfrequency. Of course, this is only one type of high-frequency limitation. In some cases the cable can be viewed as a capacitance across the signal leads. In other cases the impedance mismatch at both ends of the cable produces reflections that greatly alter the true signal.

2.2 TEMPERATURE TRANSDUCERS

The two types of transducer in general use for measuring temperature are thermocouples and resistive sensors. Thermocouples are commonly made of Chromel*-Alumel*, iron-Constantin*, copper-Constantin, Chromel-Constantin, or platinum–platinum/rhodium alloys. Resistive sensors are often made of platinum, nickel, nickel alloys, or from semiconductor materials. The choice of materials is related to accuracy, temperature range, and environmental conditions.

*A registered trademark of Hoskins Mfg. Co., Detroit, Michigan.

2.2.1 Thermocouples

A thermocouple is formed by joining two dissimilar metals at the sensing point. The two metals (wires) are then returned to a reference temperature junction. The voltage sensed at the reference junction is a measure of the temperature difference between the sense junction and the reference junction. Figure 2.5 shows a typical thermocouple arrangement. The junction of the two dissimilar metals can be made inside a temperature probe supplied by a manufacturer or formed by the user at the point of measurement.

2.2.2 Fast Response

Speed of response often requires that a thermocouple junction be electrically connected to the point being measured. Welding or brazing can be used to make this connection. In applications where this is impractical, the bond can be formed by peening the two metals into a retaining port.

Thermocouple signals rarely require a great deal of bandwidth. Because general-purpose instruments often possess excess bandwidth, a comment regarding bonding or unbonding of thermocouples is appropriate. This bond ties the input signal to an input ground. The rules given in Chapter 1 require that the shield of the input leads also be connected to this same ground. If these rules are not followed, excess noise can result that is apt to contain frequency content at 60 Hz and above. If the amplified signal is filtered, this noise may not be a problem.

Filters in instrumentation amplifiers are usually postfilters. This means that both signal and noise are amplified prior to any filtering. If the noise is excessive, it can overload the amplifier. A filtered version of an overloaded signal can be very misleading. This makes it necessary to check signal-to-noise ratios before filtering to avoid this problem. The proper treatment of shields and grounds also helps to eliminate this class of problem.

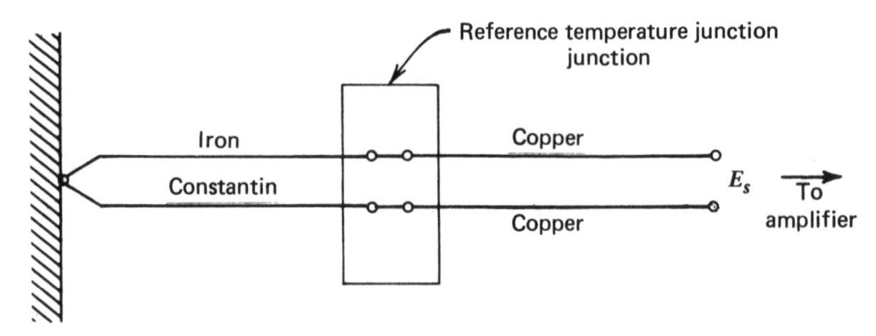

FIGURE 2.5 A typical thermocouple circuit.

In some applications grounded thermocouples are prohibited. Instrumentation amplifiers vary in their ability to cope with a floating source. A problem occurs because small input currents that must flow in the input leads require a return path to output ground. When this path is interrupted, the amplifier may malfunction. This is one form of malfunction that is difficult to observe; that is, the input signal lines may rise in voltage to the common-mode limit of the input stages. At this limit, some part of the amplifier circuitry provides a current path between the input and the output ground. Although the amplifier may appear to function satisfactorily, in reality poor performance is the rule. Again, if the user does not know the answer, the answer obtained must be accepted as correct.

2.2.3 Leakage Paths

Instrumentation amplifiers are usually designed with a leakage path (on the order of 10 megohms or more) between input guard shield and output ground. This path allows the user to operate with ungrounded transducers. Of course, if the guard shield is not properly connected, this leakage path cannot be used. The treatment of input lines where the input is ungrounded is covered in Chapter 3.

The lead-in wires from the thermocouple junction to the reference junction can vary in length from a few feet to hundreds of feet. The resistances on the two leads are not equal because the material is not the same. This causes an unbalanced input situation for the instrument amplifier. As an example, the resistances in a copper-Constantin thermocouple might be 300 ohms and 10 ohms, respectively. This resistive difference causes the conversion of common-mode noise to input signal. (See Chapter 7, under Common-Mode Considerations.)

It is often impractical to shield thermocouple leads near the point of measurement. The best alternative is to bring back one ground conductor from the test structure and tie this conductor to the shield at the earliest practical point.

Typical thermocouple signal sensitivities range from 30 to 80 μV/°C. High-temperature thermocouples often have even lower sensitivities. To increase sensitivity thermopiles can be used. A thermopile is simply a group of thermocouples in series. Each thermocouple must be brought back to the reference temperature junction to properly operate the thermopile.

Heat flow in a plate can be measured by measuring the temperature differences across the plate. If two thermocouples are used, their difference voltage represents this heat flow. This technique requires instrumentation gains in excess of 10,000.

2.3 RESISTIVE TRANSDUCERS

Thermometers that rely on temperature-sensitive resistors are often called RTDs or RTTs. These acronyms stand for resistive temperature device or transducer. These resistors can be formed as a wire grid, as a thin film, or as a semiconductor chip. The physical shape, the housing, as so on, depend on the application. For example, a long metal probe might be used to monitor a liquid, whereas a thin patch might be used to measure a surface temperature.

The measurement of resistance requries a current flow with a known current. The resulting voltage drop measures the resistance. It is important that the temperature rise due to the measuring current be small or errors can result. Obviously, the thermal resistance between the gages and any structure determines the maximum measurement current permitted.

The change in resistance over the useful range of an RTD device is often small. The bridge circuit shown in Figure 2.1 is ideal for measuring these small resistance changes. Obviously, the bridge excitation level can be set to avoid any overheating of the element.

RTDs are available with the resistive element floating or connected to its protective cover or sheath. If the element is connected to its cover, the signal is grounded when the cover is grounded. The bridge configuration in Figure 3.3 permits this one-point ground. The comments made earlier about the capacitance of a gage element to the measured structure are also valid for RTDs. A properly grounded RTD can keep system noise to a minimum.

RTD resistances can vary from a few ohms to about 10,000 ohms. High-valued resistances require the use of very fine wire. Normally, instrumentation amplifiers perform best with source resistances below 1 kohm. When

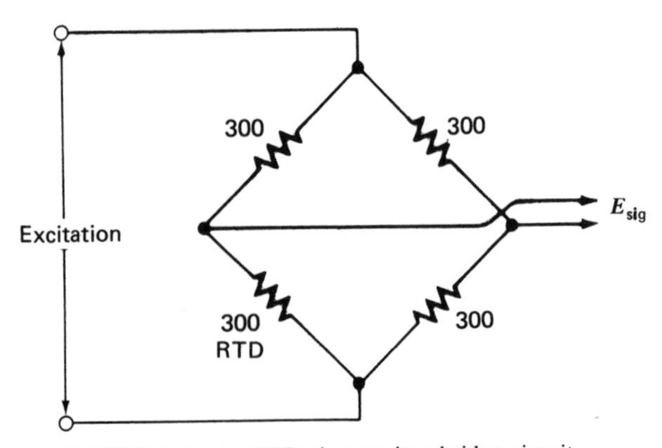

FIGURE 2.6 An RTD element in a bridge circuit.

the RTD is one element of a bridge it is not necessary that all resistor values be of equal value. One possible RTD circuit is shown in Figure 2.6.

2.4 POTENTIOMETERS

Mechanical motion can be measured by using a linear potentiometer. Displacement causes motion of the slider that is then sensed as a voltage. If the potentiometer is used as two arms of a bridge, the null point can represent a reference position. Displacements around this null point produce plus and minus signals (see Figure 2.7). Signals from potentiometers are usually high level.

2.5 PRESSURE TRANSDUCERS

Pressure transducers often use a diaphragm that strains under pressure. This diaphragm may be a flat or corrugated round disk. Strain can be detected as mechanical motion, capacitance change, magnetically, or by strain gages placed on the diaphragm. Four active arms are frequently used to improve signal amplitude and linearity.

Pressure gages are available with internal electronics that include an excitation source, an amplifier, and any conditioning required to zero, linearize, or set full scale for the output signal. Full-scale output signals are typically 0–5 V. These signals can often be recorded or displayed without further processing. If the excitation source is sufficiently well isolated, the output signal can be grounded at the recording point without creating a ground loop or adding excessive noise. Users, however, frequently process a 0–5 V pressure signal through an instrumentation amplifier just to insure a noise-free system. In this case, the excitation source can be grounded and the instrumentation can be used to attenuate common-mode signals (ground differences of potential). Also the instrumentation provides for offsets, filtering, gain, and so on.

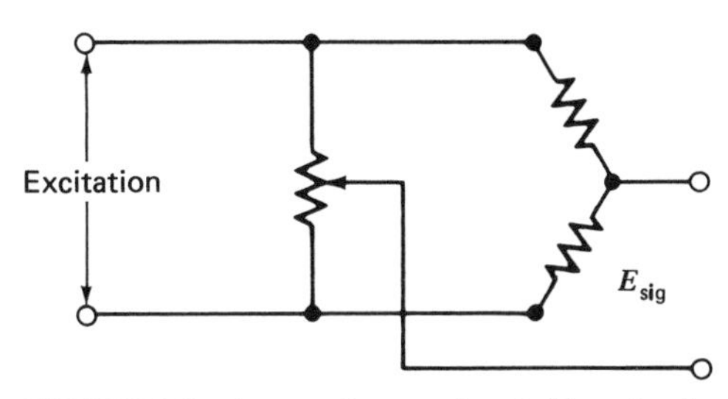

FIGURE 2.7 A potentiometer in a bridge circuit.

2.6 LOAD CELLS

Load cells or force transducers are used to measure the force between two parts on a structure. The force causes a small mechanical deformation that can be sensed optically, capacitatively, magnetically, or with strain gages. Load cells can be used to measure thrust, weight, or torque. Typical strain gage load cell designs use two or four active arms.

When strain gages are used, the manufacturer usually conditions the signal to be linear and provides a known output voltage per unit force for a given excitation level. It is usually up to the user to provide gage excitation and signal amplification for the transducer.

The gage elements float from the transducer housing, but the capacitance to the housing can be a source of noise. Later comments regarding noise and rf pickup are applicable.

2.7 PIEZOELECTRIC TRANSDUCERS

The piezoelectric effect is used in transducers to measure acceleration and shock. When a force is applied between the faces of certain crystals, a charge will appear. Conversely, if a charge is introduced, a strain will result in the crystal.

Many "crystal" materials can be used, but the most frequently used material is barium titanate. Each manufacturer adds one's own set of impurities to produce a proprietary product. Different base materials allow for higher operating temperatures. For example, barium titanate is limited by its curie point to 120°C, whereas lead metaniobate has a curie point of 570°. To increase the output signal level, crystal elements are often placed in series. The method of stressing the crystals is another variable used by the manufacturer to improve senstitivty.

The internal resistance shunting an element is often above 10^9 ohms. This resistance places a limit on the low-frequency response of the transducer. From a few hertz to perhaps 30 kHz the transducer can be considered a capacitance in series with a voltage generator. The leakage resistance shunts this series combination. (see Figure 2.8). The leakage resistance of the cable

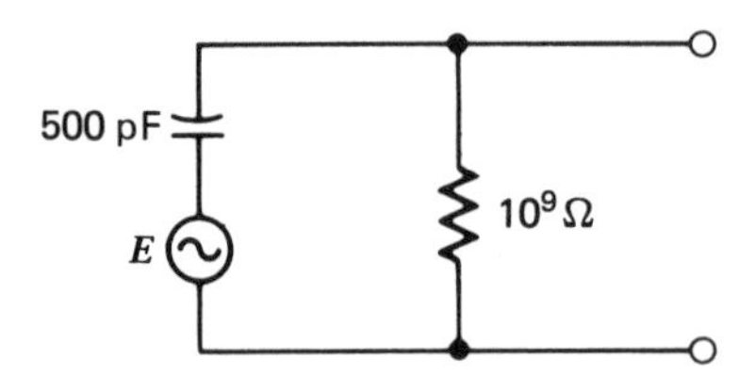

FIGURE 2.8 A crystal accelerometer equivalent circuit.

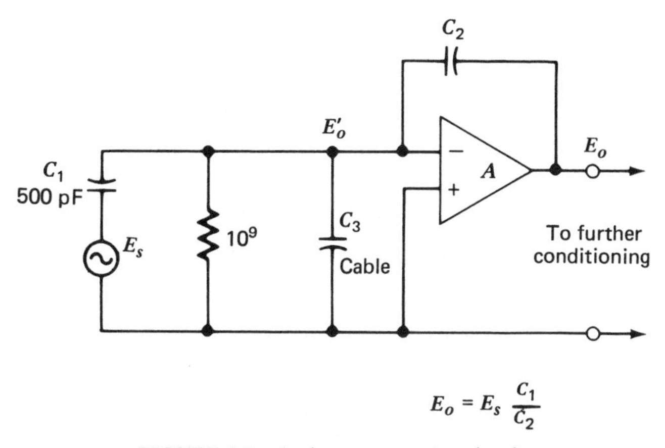

$$E_o = E_s \frac{C_1}{C_2}$$

FIGURE 2.9 A charge converter circuit.

and the input impedance of any subsequent amplifier are in parallel with this internal resistance. A typical 500 pF transducer with 10^9 ohms leakage resistance would have a low frequency -3 dB point at 3 Hz.

When voltage amplification is used, the cable capacitance parallels the transducer capacitance and this attenuates the available signal. Calibration requires that the cable capacitance be accurately known. To circumvent this difficulty, charge amplifiers are generally used. The charge amplifier employs a charge-converter input stage. The charge appearing on the transducer is converted to a voltage at the output of the converter. This circuit is shown in Figure 2.9.

The gain for voltage E_s is simply the ratio C_1/C_2. The effect of cable capacitance C_3 is to reduce the signal-to-noise ratio and to reduce the loop gain available to keep the signal gain at the value C_1/C_2.

It is interesting to note that the voltage at E_1 is near zero for A large. The voltage E_s gives rise to a charge $Q_1 = C_1E_s$. This charge must reside on C_2 to force E^1 to zero. The voltage drop across C_2 is the output voltage. This voltage is simply

$$\frac{(Q_1)}{C_2} = \frac{(C_1E_s)}{C_2} = \left(\frac{C_1}{C_2}\right) E_2$$

or the gain of the converter times E_s.

The mass of the transducer should ideally be zero, for this mass changes the force required to accelerate the structure. Any materials used to isolate the transducer from its element or the transducer case from the structure add to this mass. At high frequencies an insulating barrier can also act as an added dynamic element that can obscure the measurement at hand. For these reasons it is often desirable to electrically connect one side of the

FIGURE 2.10 An industrial transducer. (Photograph courtesy of Endevco Corporation, Capistrano, Calif.)

transducer to its case. This can be tolerated only if the instrumentation can accept a grounded and unbalanced source. The problems this presents are covered in subsequent chapters. A photograph of an industrial transducer is shown in Figure 2.10.

CHAPTER THREE

TRANSDUCER CONNECTIONS

"A fine cage won't feed the bird."

3.1 Strain Gage Bridges

Gage elements are most frequently bonded onto metal structures. These structures are usually grounded for safety reasons. If the structure is operational, there are often electrical gradients along the structure caused by various drain currents. These currents can arise from motors, controllers, relays, safety wires, or external rf fields. The larger the system under test the greater the ground potential differences can be.

As discussed earlier, one, two, or four active arms may be used to measure a single parameter. Active elements are usually mounted close together so that the surface gradient for one gage grouping is rarely a problem. The capacitance from one gage element to the structure can easily exceed 100 pF. If the structure is not used as the zero reference conductor, parasitic currents can flow in the gage resistance through this capacitance. This effect is shown in Figure 3.1. The potential difference between grounds **1** and **4** causes current to circulate in loop **1–2–3–4–1** and this current flows in element R_4. High-frequency pickup can cause rectification problems in the amplifier. Once an overloading signal is sensed in R_2 it can only be removed prior to

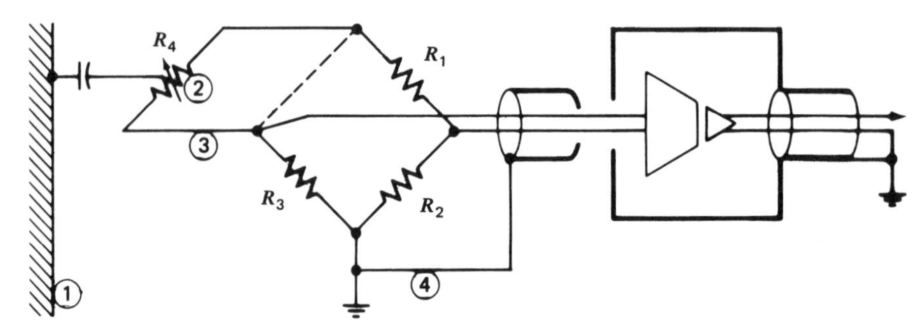

FIGURE 3.1 The parasitic problem for one active strain gage element.

amplification at the amplifier input. When two or four arms are active, multiple capacitances to the structure can also couple common-mode signals into the amplifier. Amplifiers are often incapable of handling high-frequency common-mode signals and the comments made previously are applicable. (see Section 7.4).

The bridge completion resistors R_1, R_2, R_3 can be located at or near the test structure **1**, in a separate conditioning module, or within the instrumentation module. In any of these three locations, the bridge arms need to be shielded by the proper zero-signal reference potential. To avoid the parasitic problems just discussed the input shield should connect to **1** and not to a ground defined by the excitation source.

The excitation source can be located in a separate module, that is, in a module common to many gages or within the instrumentation module. The excitation source should also be shielded by the proper zero-signal reference potential.

A common excitation source can be used when the electrical gradient along the structure is known to be low. The problem that can arise when a gradient exists is shown in Figure 3.2. The potential difference E_{16} drives current through capacitance C_{12} in the loop **1–2–3–4–5–6–1**. Note that there is no current driven through capacitance C_{67}. This unwanted noise current in the gage elements develops a voltage that must be processed by the instrument amplifier. If common supplies are employed, the user should be careful to test for input noise contributed by electrical gradients on the structure.

Common supplies can create a reliability problem. If the supply is affected by a short to a ground of one gage, all signals may be disturbed. If the common supply is floating, a ground on one gage may not be detected except as a possible change in noise in the system. The gage that shorts to ground is obviously suspect, but if the system shows no signs of distress, the problem

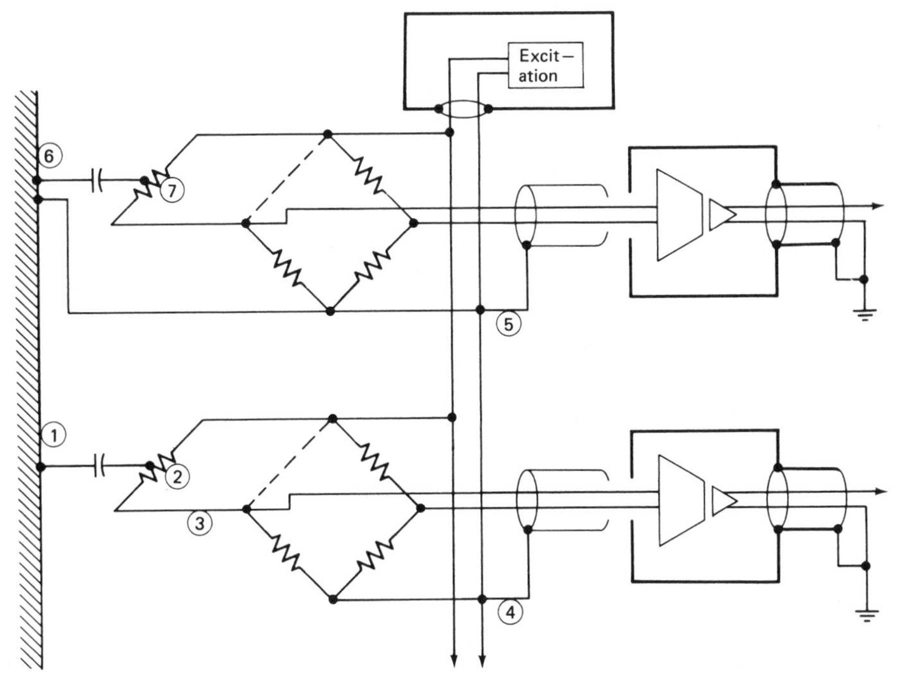

FIGURE 3.2 The common excitation problem.

may go undetected. If two gages short to ground, then both gages are thrown out of service or the supply is affected. This reliability problem is the main reason why most users specify separate isolated supplies for each gage.

3.1.1 Bridge Excitation—Constant Voltage

When strain gage bridges are separately excited, remote voltage sensing can be used to regulate the excitation level at the gage elements. A 100-ohm bridge with 10 V excitation requires 100 mA of excitation current. Each ohm of lead wire reduces the excitation level by 0.1 V. One hundred feet of #26 copper wire has a loop resistance of 8 ohms. Remote voltage sensing can work for two or four active arms. Bridge completion resistors located in a conditioning module or in an amplifier will be excited by a voltage higher than the sense voltage, but this does not create a signal error.

Remote sensing of excitation level is optional when one supply is used to excite multiple gages. When sense leads are used they draw some current, but this is usually specified to be in the microampere range. The sense leads are usually run in parallel with the excitation leads inside a common shield.

3.1.2 Bridge Excitation—Constant Current

Constant currrent sources can be used to avoid the need for remote voltage sensing. When the source current is constant, the voltage developed across the full bridge is independent of cable length. When bridge completion resistors are located in a conditioning module or in an instrument amplifier, this part of the excitation load becomes the power supply source impedance. Note that an ideal current source has an infinite source impedance. However, unless the completion resistors are located near the active arms, the benefits of a constant current source are lost.

Constant current excitation improves signal linearity for many of the often used bridge configurations. In fact, some configurations become exactly linear for any ΔR. Signal linearity for various bridge configurations is shown in Table 2.1.

3.1.3 Calibration Processes

Three methods of electrical calibration are possible. The first method adds a known voltage in series, with the input leads internal to the amplifier. This method is rarely used because a proper treatment is expensive and difficult to implement. The second method uses a relay to disconnect the bridge and substitute a known voltage* within the amplifier. This method might calibrate the amplifier, but it avoids calibrating the bridge proper. The third method places shunt resistors across the bridge arms to simulate strain values. It is usually the practice to shunt only one bridge arm at a time although one, two, or four arms may be active.

Calibration resistors are usually high in resistance to simulate a strain. Typical values vary from 10 k to 100,000 ohms. These resistors are switched by relay contacts or solid-state switches located within a conditioning module or within the instrumentation amplifier. Ideally these resistors should be shunted across the active arms of the bridge rather than across any completion resistors located away from the structure under test.

Accurate calibration requires that separate leads be run to the active arms to handle the shunting process. If this is not done, the IR voltage drop in excitation leads will be in series with the calibration resistor. This can introduce a 1 or 2% error when the excitation leads have a resistance of a few ohms. Calibration currents are obviously much smaller than excitation currents. For this reason one of the calibration leads can share a signal line. This option is shown in Figure 3.3. Resistor R_2 is shown going back to the

*The known voltage can be the excitation source attenuated by the resistance bridge and a series resistor. See section 6.8.1.

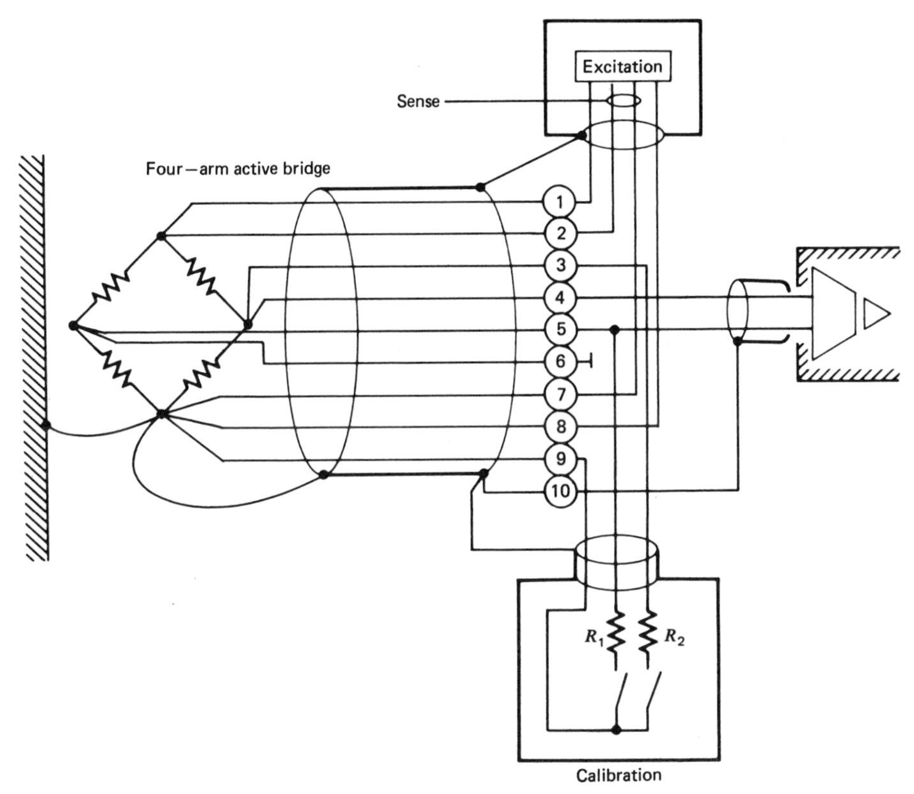

FIGURE 3.3 Connections in a calibration and remote sense circuit.

bridge on conductor **3** and resistor R_1 is shown sharing conductor **5**, one of the signal conductors. Both techniques are viable.

The excitation source, calibration circuitry, and instrumentation amplifier can be packaged separately or in combination. The details of control for relays or the power for the excitation supply are not shown in Figure 3.3. In general, relay power and logic control circuits can violate the electrostatic shield that surrounds this input circuitry. Therefore some means of isolation is required. These techniques are discussed in Chapter 6.

3.1.4 Floating Excitation Sources

In the preceding sections the excitation source was shown grounded to the structure. Another technique that is sometimes specified requires that the excitation supply float and that one of the signal lines be grounded. This technique provides a balanced excitation but unbalances the input to the amplifier. This configuration is shown in Figure 3.4. One input lead sees the bridge impedance, whereas the other lead sees ground. This technique finds

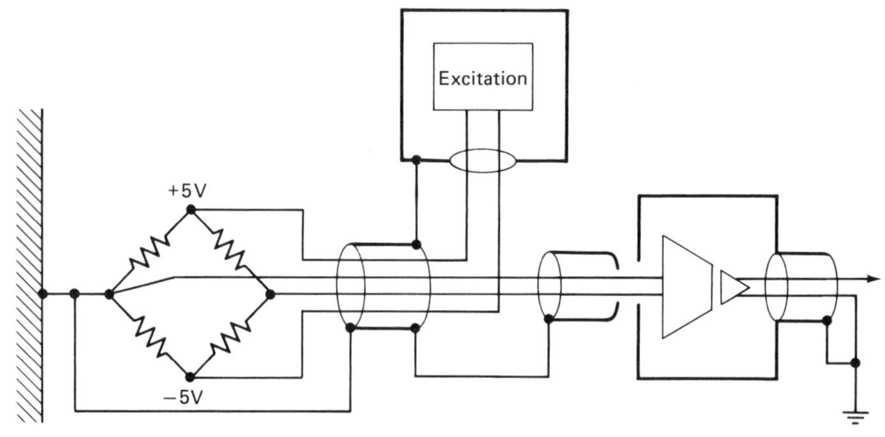

FIGURE 3.4 A floating excitation configuration.

its origin in the days of single-ended amplifiers where the output of the amplifier had to be grounded and the gage was necessarily insulated from the structure. A floating supply was a necessity in this arrangement.

A floating power supply was discussed in Chapter 1. This supply must be carefully designed to avoid unwanted current flow in the bridge arms. If this noise current is to contribute signals below typical amplifier noise levels, then the noise current must be kept below 1 or 2 nA rms. For 117 V at 60 Hz the mutual capacitance between the primary coil and the secondary shield must be held to less than 0.01 pF—a difficult feat.

3.1.5 Remote Bridge Sensing

Remote sensing usually implies monitoring the excitation line at a remote point. These sense lines carry very little current compared with the excitation lines. When one active arm is used, bridge current must be carried to this one element. Here, too, signal sensing should not involve leads that carry bridge current. To solve this problem a separate signal line needs to be returned to the amplifier.

Bridge completion can take place in the instrumentation or remotely near the active elements. If the distance from the completion point to the active element is short, then special signal-sensing lines are unnecessary. The one active arm-sensing problem is shown in Figure 3.5. Here line **1–2** is used to sense one side of the signal, thus eliminating the IR drop in line **1–3**. Note that there is no provision to reduce the effect of the IR drop in lead **4–5**. A separate line **4–6** can be used to calibrate the one active arm, which removes the error caused by the IR drop in lead **4–5**.

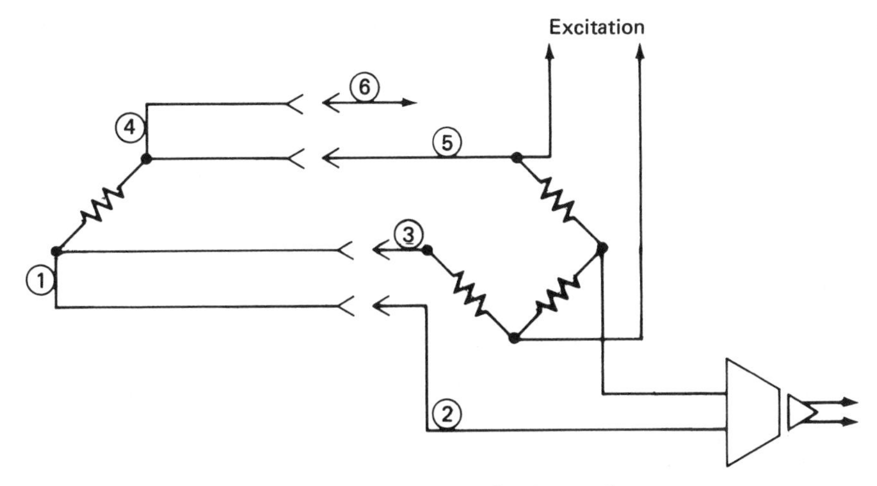

FIGURE 3.5 A separate signal sense line.

The many possible configurations include one, two, four active elements, remote or local bridge completion, remote or local excitation sensing, remote or local signal sensing, and remote or local calibration connections. The user must be aware of the errors caused by IR drops in the excitation leads, including temperature effects. With this knowledge the user can select a wiring scheme that is best suited to the applications. Present-day instrumentation practices allow users to connect up to ten wires to their gage. This is sufficient to handle the worst case. Arrangements at the input connector define which sense or calibration leads are used.

3.2 THERMOCOUPLES

Thermocouples can be used to measure fluid temperatures, heat flow, surface temperatures, and so on. Temperature rise around a ballistic impact or after an explosion requires a high-speed measurement. The temperature in industrial processes or in a fuel tank requires very little bandwidth. Fast response requires low thermal mass and intimate contact with the surface being measured. Fluid measurements often require a large surface area to get an average reading. Bonding the transducer to the measured surface obviously improves the speed of response.

If a differential device is used to amplify a thermocouple signal, a source ground is recommended to define the input common-mode level. This ground point should be at the point of measurement to avoid electrical gradients on the surface. It is often difficult to obtain shielded thermocouple wire and thus connections to the reference junction are often run unshielded. If this run is short, the source resistance unbalance will be low and common-mode conversion to differential mode will not be a problem. Also note that external grounds near the test are less contaminating than grounds far removed from the test. This is shown in Figure 3.6.

Reference junctions are best located near tests for the preceding reasons. Because one reference junction is often used for many signals, it has a parasitic association with a local ground unless a guard plate is provided. This guard should be tied to an average point on the structure. The reason this is critical is related to the postfiltering problem discussed in Section 2.1.2. Overloading prior to filtering can lead to strange results. Reference junctions can be ice and water, electrically regulated heat plates, or the vaper point of liquid nitrogen.

Differential temperature measurements do not require using a reference junction. The user needs only to know the approximate temperature so that the appropriate voltage coefficients are used for the particular thermocouple materials. Thermopiles, on the other hand, require that each thermocouple placed in series makes a connection to the same reference junction.

When thermocouples are not bonded to the structure, a ground connection along a thermocouple lead introduces a spurious thermocouple. To avoid this problem a Wagner ground is used. This configuration is shown in Figure 3.7. This arrangement serves two purposes. It balances the input leads with respect to the shield and defines the shield at the source ground potential.

Although the Wagner ground divider is best located near the reference junction, it can also be located in the instrument. The key issue is whether the shields are referenced to a local unknown potential or to the structure under test as shown in Figure 3.7. The resistor value R should be about 1000

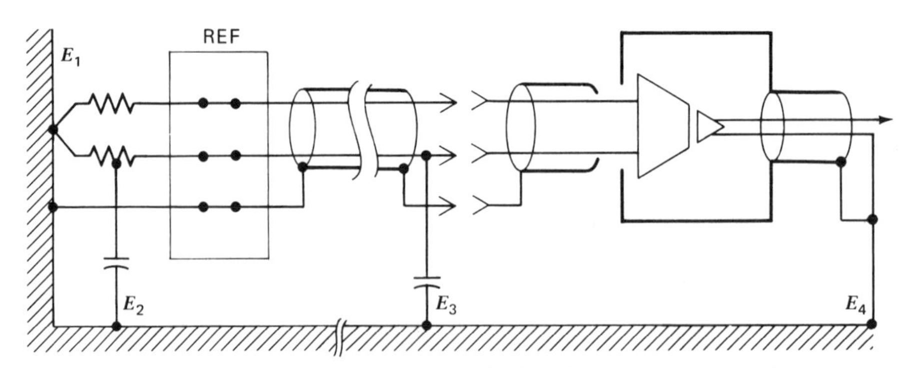

FIGURE 3.6 The gradient of voltage contamination along a thermocouple cable run.

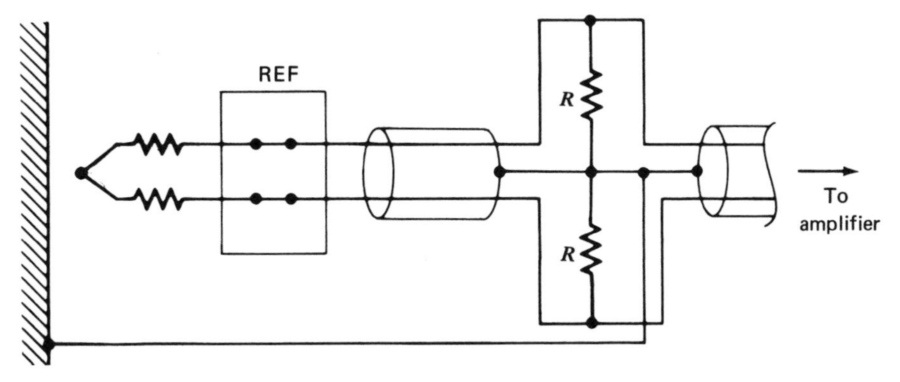

FIGURE 3.7 A Wagner ground.

ohms. This is high enough to avoid attenuating the signal and low enough to allow good performance from the amplifier.

Input filtering is desirable where thermocouple signals are not properly shielded. This filtering helps solve the problem of amplifier overload. Filtering can be located with the Wagner ground shown in Figure 3.7 or placed within the instrument as part of its signal conditioning. Figure 3.8 shows a balanced input filter circuit. The value of R_1 should not exceed 1000 ohms and C can be a polar capacitor. Note that the leakage resistance of a polar capacitor is low for small bipolar signals. For $C = 10\ \mu F$ and $R_1 = 1k$, the input filter has a -3 dB point of about 10 Hz.

The filter in Figure 3.8 does not reduce the common-mode level. If the Wagner ground is selected as the output ground, the common-mode signal

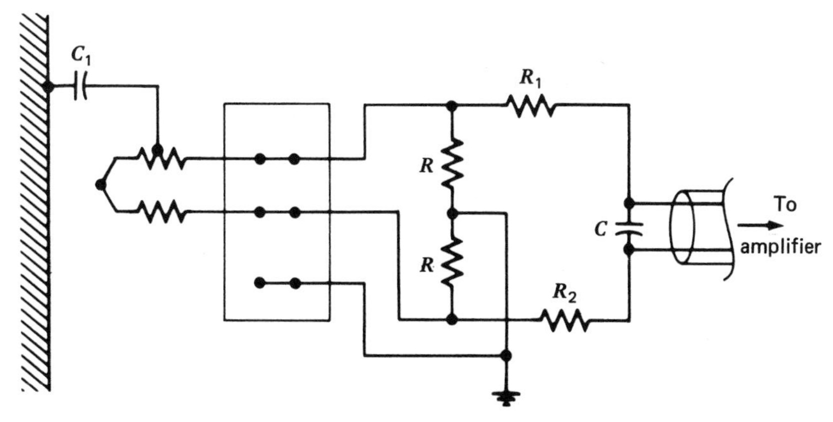

FIGURE 3.8 A balanced input filter.

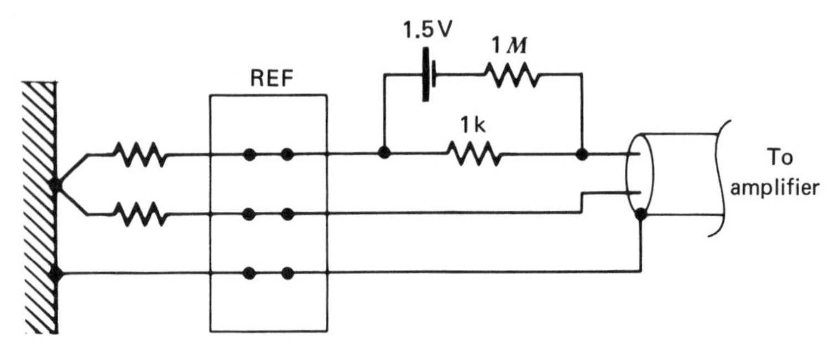

FIGURE 3.9 A battery RTI offset.

will be attenuated. In this arrangement parasitic differential currents coupled by capacitance C_1 are attenuated by the input filter.

Reference junctions should operate in the same temperature range as the temperature measurement. If the difference in temperature is excessive, the range of signal variation will be a small percentage of the developed signal. This in turn limits the maximum gain setting of any subsequent amplifier.

One approach used to avoid this difficulty is to provide a referred-to-input (RTI) offset in the instrument. Because this is not an easy task electronically, some users have resorted to small battery attenuators. The offsets are usually in the millivolt range and battery shelf life can be realized. A typical circuit is shown in Figure 3.9. This circuit offsets the thermocouple signal by 1.5 mV. See Section 7.7 for further treatment.

In situ calibration of thermocouples is not too practical. Voltage substitutions or voltage insertion are possible solutions. If the battery in Figure 3.9 were opened, the resulting 1.5 mV step voltage could serve as a one-point calibration. RTDs are usually treated as one arm of a Wheatstone Bridge. Two RTDs can be used to measure temperature differences or to increase signal levels. These circuits are nearly identical to strain gage configurations discussed in Section 3.1. The treatments of bridge completion arms, excitation sources, and so on are all applicable.

3.3 POTENTIOMETERS

Potentiometers of the resistance-slider type are available in a variety of configurations. Multiturn potentiometers can be mechanically coupled to provide excellent resolution. The output signal is usually high in value and therefore little or no amplification is required. Common-mode difficulties, however,

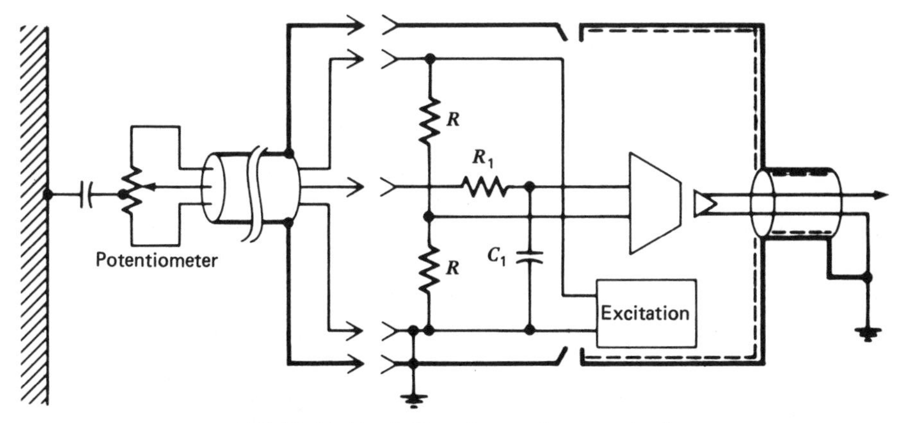

FIGURE 3.10 A filtered potentiometer circuit.

can still exist because common-mode levels are a function of the system and not of the transducer output.

Mechanically linked potentiometers are by their very nature bandwidth limited. For this reason, filtering to output ground can be used to limit amplifier performance and remove unwanted common-mode signals. This connection is shown in Figure 3.10. This circuit obviously poses little difficulty for the instrument amplifier. The completion Rs can be located in a conditioner unit or within the instrument. The value of R_1 should not exceed 10 kohms or a value allowed by the manufacturer. Note that R_1 unbalances the amplifier. A filter can be placed in both input lines to maintain a balanced source if this is recommended.

The currents drained by the filter in Figure 3.10 flow inside the amplifier. Current drained by the input shield also drains inside the amplifier. If serious rf levels are expected, these currents should drain through a ground conductor that skirts the amplifier. This arrangement is shown in Figure 3.11.

3.4 PIEZOELECTRIC TRANSDUCERS

Piezoelectric transducers are usually capacitive single-ended sources. The problem of floating these transducers is discussed in Section 3.1.3. The input cable must be a low-noise and have low leakage capacitance. No part of the input lead can be bared. If the transducer capacitance is 1000 pF and the RTI noise is 10μV, a nearby ground differing in potential by IV must be attenuated by a factor of 100 dB. This is an effective mutual capacitance of 0.01 pF.

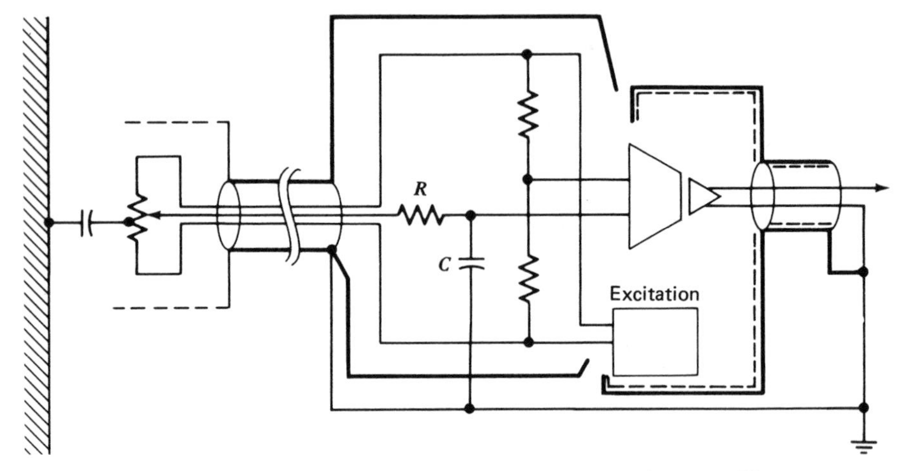

FIGURE 3.11 The drain of input shield current around the amplifier.

3.4.1 Charge Converter

The charge converter circuit shown in Figure 2.9 is a single-ended device. The conditioning amplifier that follows the converter can be single-ended or differential. The differential amplifier provides the high input impedance required for common-mode rejection (see Section 1.6). The charge converter input impedance is a source of line unbalance and can convert common-mode signals to differential mode. If the value of C_2 in Figure 2.9 is 1000 pF, the source impedance at 60 Hz might be 1000 ohms. A differential transducer or a dummy capacitor is required to balance the input circuit. Balancing has meaning only for a grounded transducer used with a differential amplifier.

A coaxial cable between the transducer and the charge converter forces the outer conductor to serve both as a signal conductor and as a shield. If the input cable runs any distance, reactive noise current flowing in this outer conductor can be converted to differential or input signal. If rf pickup is a problem, a triaxial cable should be considered. The outer shield should be connected to ground only on one end to avoid a low impedance loop. The coaxial cable must be insulated at each bulkhead or interface. Note that one outer shield can serve several input cables. The single triaxial shield is treated in Figure 3.12.

3.4.2 Voltage Amplifiers

Voltage amplifiers are often used to amplify signals from piezoelectric transducers. The disadvantage is the attenuation of signal as a function of cable

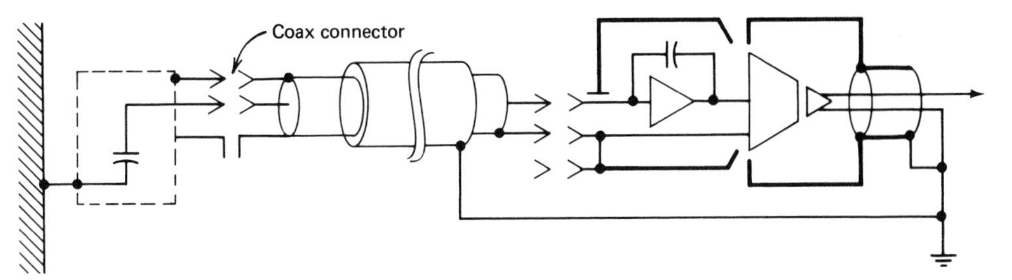

FIGURE 3.12 A triaxial shield used with a piezoelectric transducer.

length. It is interesting to note that as cable length increases, the signal-to-noise ratio is reduced equally for both charge- and voltage-type instruments. Cable lengths need to be standardized if signal attenuation is to be accurately known.

Voltage amplifiers are just as sensitive to input cable noise as charge amplifiers. However, modes of noise coupling are not the same. Mechanical changes in capacitance can introduce noise into a voltage amplifier, whereas a charge amplifier may be unaffected. Also, charges released in the cable due to friction can introduce noise into a charge amplifier but may have negligible effect on a voltage amplifier.

A very high impedance is needed in a voltage-type amplifier. If the -3 dB point is to be 1 Hz, the input Z must be greater than 200 megohms. This impedance must remain high for all signal levels. It is common practice to use a coaxial cable and miniature connectors at every interface to keep external noise pickup to a minimum. The triaxial cable discussed in Section 4.9 and shown in Figure 3.12 is applicable.

The high input impedance of a typical transducer represents a significant line unbalance. Differential amplifiers cannot function well when connected to capacitive sources, but even if they could operate properly, the line unbalance problem would preclude any reasonable common-mode rejection ratio. However, a postdifferential amplifier similar to that shown in Figure 3.12 can be used.

3.4.3 Preamplifiers

The best signal-to-noise ratios for piezoelectric amplifiers result with short input cables. If a preamplifier can be placed near the transducer, then the long cable run can be driven at high signal levels, thus avoiding the cable problem. The electronics positioned near the transducer must be able to withstand the transducer environment.

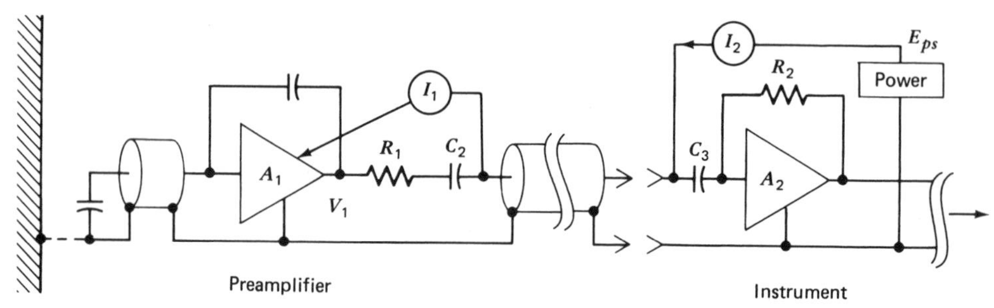

FIGURE 3.13 A charge preamplifier system.

Several techniques have been developed where the two-wire cable between the preamplifier and the receiver carries both power and signal. Because the signal is at ac and power is at dc this separation is practical. This arrangement can work with either a voltage or charge preamplifier. Figure 3.13 shows a charge-converter circuit where amplifier A_1 converts charge on C_1 to a signal voltage V_1. Resistor R_1 connects this signal to the summing point of amplifier A_2. Resistors R_1 and R_2 determine the gain of A_2. Capcitors C_2 and C_3 are made large enough to couple the lowest frequency of interest. The cable run is a low impedance at ac. The average dc value of the cable is determined by current sources I_2 and I_1 that convert power E_{ps} to voltage E that is used to operate charge converter A_1. Making the cable the summing point for amplifier A_1 means that amplifier A_1 need not drive the cable from a voltage source and supply large reactive currents. The electronics that follows A_2 can be single-ended or differential.

The preamplifier section shown in Figure 3.12 can also contain an active filter. The filter section would follow A_1 and would in turn drive resistor R_1.

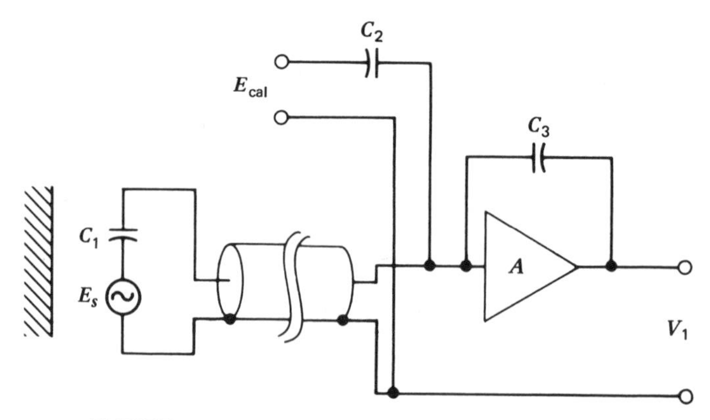

FIGURE 3.14 A charge converter calibration scheme.

A prefilter can remove noise that would otherwise overload subsequent gain stages. It is obviously futile to filter signals at the output after they have already exceeded full scale.

Charge amplifiers can be easily calibrated without switching the input signal lines. The method is shown in Figure 3.14. Capacitor C_2 converts E_{cal} to a charge that results in voltage V_1. If E_s generates a full-scale charge on C_1, then E_{cal} can be ganged to the full-scale charge selector switch of the instrument. When a preamplifier system is used, this form of calibration is not practical.

CHAPTER FOUR

PLANT ENVIRONMENT

"He who keeps off the ice will not slip through."

Plant environment frequently has an impact on the signals being processed. Cables in open trenches are influenced by the effects of ground current. Cables in raceways encounter much different environment. This chapter treats a few of the problems experienced by signal lines in typical installations.

The impedance along a metal surface is a function of frequency, thickness, and type of material. A voltage gradient can result from reflecting radiated energy or from conducted control or power currents.

The ground plane is sometimes a grid of reenforcing bars, a sheet of aluminum, a raceway, or simply an open terrain. Each of these surfaces has different electrical characteristics. Frequently, cables will traverse several ground plane regions and potential differences result from multiple sources.

The ground potential difference resulting from ground current is one source of common-mode signal. A second important source is from the cable ground-loop area. Radiated fields that cross this loop perpendicular to the cable run couple voltage to the loop that is roughly proportional to the loop area.

Another coupling mechanism that must be considered is the conversion of shield current to differential-mode signal. This shield current can result from the ground-loop coupling described above or from capacitances to nearby grounds.

4.1 GROUND IMPEDANCES

Earth currents can vary significantly, depending on locality. Directly under transmission lines the 60 Hz current can be many amperes. The earth carries any power line unbalance current and serves to drain currents from filters and ground capacitances. The condition of the earth varies, depending on weather and soil conditions. For these reasons it is difficult to assess the frequency spectrum and magnitude of ground potential differences ahead of time. Voltages in excess of 10 V are often encountered.

Within a structure, metal ground planes provide a very low impedance even if the plane is only 0.03 mm thick. Some typical impedances are shown in Table 4.1.

4.1.1 Impedance of Ground Straps

Conductors are often used to carry a ground to a structure. This impedance has both a resistive and an inductive reactive component. At high frequencies the reactive component dominates and heavier wire serves no useful purpose. Table 4.2 shows a few typical impedance values for solid copper wire.

Ground straps often serve both as a safety conductor and as an rf drain. To reduce the strap inductance, rectangular conductors can be somewhat more effective.

TABLE 4.1
Metal Plane Inpedance in Ohms per Square[a]

	Copper Thickness (mm)			Iron Thickness (mm)		
Frequency	0.03	0.1	10	0.03	0.1	10
50 Hz	0.57	0.17	0.017	3.4	1.0	0.03
500 Hz	0.57	0.17	0.017	3.4	1.0	0.27
10 kHz	0.57	0.17	0.030	3.4	1.2	1.2
1 MHz	0.60	0.28	0.37	9.5	11.2	12
10 MHz	0.92	0.95	1.2	35	38	40
100 MHz	3.0	3.5	3.7	121	125	126

[a]Values are all in milliohms.

TABLE 4.2
Impedances for Straight Copper Wires

Frequency	# 10 AWG Wire Length 1 meter	# 10 AWG Wire Length 10 meters	# 22 AWG Wire Length 1 meter	# 22 AWG Wire Length 10 meters
50 Hz	3.3 m	33 m	53 m	530 m
500 Hz	5.28 m	65 m	53 m	530 m
10 kHz	83 m	1.11 Ω	0.11 Ω	1.39 Ω
1 MHz	8.26 Ω	1.11 Ω	10 Ω	129 Ω
10 MHz	82.8 Ω	1.1 k	100 Ω	1.3 k

[a]m = milliohms, Ω = ohms, k = kilohms.

4.2 LICENSED TRANSMITTERS

Fields radiated from a variety of licensed transmitters can impact system performance. The effective radiated power levels can vary over a wide range. Figure 4.1 shows some typical values. Effective radiation includes the gain of the antenna. Note that 60 dB equals 1 MW.

The concern of systems engineers is the field strength at their location. The equation for the E field is given by

$$E = \frac{1}{D}\sqrt{30W} \tag{4.1}$$

where E = field strength in volts per meter
D = distance from transmitter in meters
W = effective radiated power in watts.

A television transmitter at 10 km has a field strength of 1 V/m.

4.3 *E* FIELD COUPLING

The E field inside a building is attenuated as a function of frequency and distance inside the building. The building steel is the main reason for this attenuation. At frequencies between 30 and 300 mHz the attentuation averages about 20 dB. For signals in the 1 MHz range, attenuations are apt to be about 60 dB.

The equations that relate to E field coupling are well defined for dimensions less than a quarter wavelength. Above this frequency, coupling can be very sensitive to physical dimensions. Cables that traverse a single region of ex-

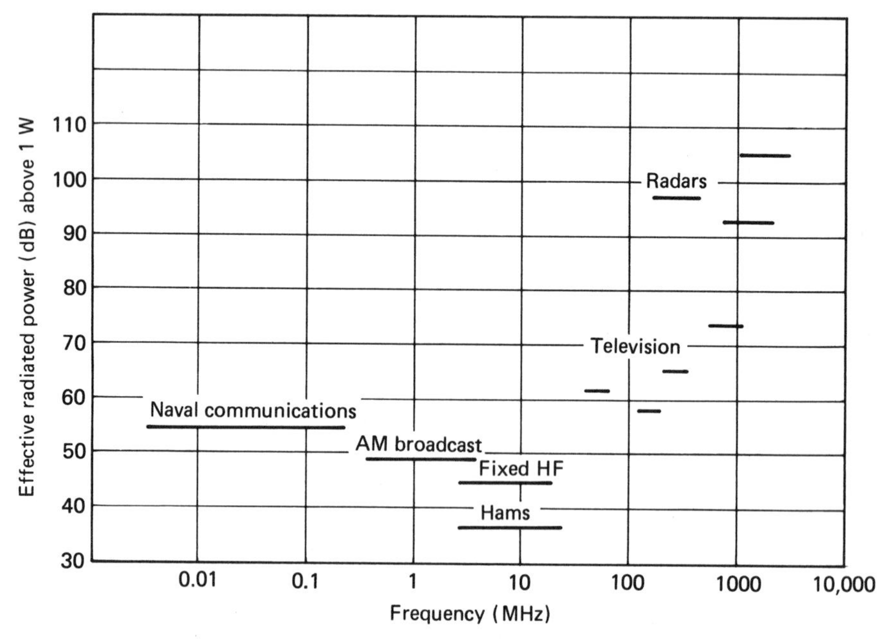

FIGURE 4.1 Typical radiation sources.

ternal radiation present a simpler problem. In practice cables in closed and grounded raceways will encounter very little external E field.

The E field ground-loop coupling for dimensions less than a quarter wavelength is given by

$$V = \frac{2d\pi hE}{\lambda} \tag{4.2}$$

where d = length of loop in meters
h = height of loop above ground plane in meters
λ = wavelength of the E field in meters
E = field strength in volts per meter
V = voltage

The direction of the radiation and its polarization are assumed to be optimum. Because these factors are rarely known, this equation gives a worst-case value.

The emf that is developed by an external radiating field is usually a common-mode signal at the instrument input. Figure 4.2 illustrates the nature of the pickup for a typical input cable. The voltage V will cause shield current in the shield-to-ground impedance located within the instrument unless an external filter is used to bypass the energy.

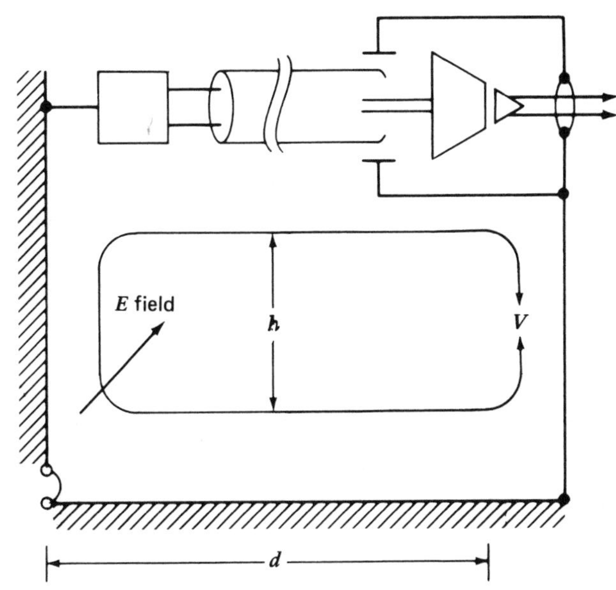

FIGURE 4.2 *E* field coupling into a cable/ground loop.

4.3.1 Earth Penetration

The area of the loop in Figure 4.2 is well defined for in-house cabling, but over open terrain the value of h is not so clearly defined. The E field penetrates the earth as a function of soil condition and frequency. The phenomenon is governed by ordinary skin effect. For very arid soil the value of h can be surprisingly large.

Table 4.3 shows a tabulation of E field penetration for various soils and for various frequencies. A doubling of attenuation occurs for each doubling of depth.

TABLE 4.3
Radiation Penetration for Various Earth Conditions[a]

	Terrain			
Frequency	Desert	Typical Land	Wetlands	Ocean
60 Hz	600 m	370 m	190 m	27 m
400 Hz	200 m	130 m	68 m	9 m
100 kHz	13 m	7 m	4.0 m	62 cm
10 MHz	1.3 m	85 cm	40 cm	6.2 cm

[a]The depths shown are for 6 dB attenuation.

4.3.2. Differential Coupling

Currents flowing in a shield can couple differential signals into internal conductors. The conversion process is complex and depends on eccentricity, type of shield, and on the twisting of internal conductors. This conversion process is apt to be greatest for coaxial cables. To place this conversion in perspective consider an RG 58/U cable. The conversion impedance is 0.01 ohm at 100 kHz per meter of cable. This is due primarily to the dc resistance of the outer jacket. At 10 MHz the conversion impedance is 0.1 ohm per meter of cable. If the shield carries 1 mA and is 100 m long, the conversion voltage is 10 mV. This is a significant voltage if the cable is the input to a charge amplifier.

The conversion impedance for various cables is best determined empirically. Generally speaking, rigid thick-walled coaxial cables will have the lowest differential coupling. The shielded twisted pair is a very close second. Note that rigid cable and low-noise cable are not necessarily one and the same.

Direct radiation on open signal leads produces a differential signal that is given by equation 4.2. The value of h is the spacing between conductors. Signals are sometimes carried over open conductors such as telephone lines. These leads are particularly susceptible to differential coupling from radiated fields.

4.4 PARALLEL CONDUCTORS

Parallel conductors can cross couple energy. This coupling is dominated by magnetic or capacitive processes, depending on conductor geometry. If the characteristic impedance of two conductors is well below 377 ohms, the coupling mechanism is mainly magnetic. The characteristic impedance is

$$Z_C = \sqrt{\frac{L}{C}} \tag{4.3}$$

where L and C are the inductance and capacitance, respectively, per unit length.

Magnetic coupling can occur on pc (printed circuit) boards, between conductors on ribbon cable, or between power conductors and shielded signal cables. Twisting pairs of wires can significantly reduce this coupling. This is not practical on pc boards or for coaxial cables.

Electric field coupling can also occur on pc boards and on ribbon cable.

This coupling can be both common mode and differential in nature. Shielded cables are not subject to direct differential electric field coupling.

Fields around pairs of conductors are symmetrical about a midplane. For the electric field this plane is an equipotential plane. Once the field is known between two parallel conductors, it is also known for the single conductor over a ground plane. This is illustrated in Figure 4.3.

The magnetic field around two current-carrying conductors is also symmetrical about a midplane. This midplane can be replaced by a conductor and the fields remain unchanged. Again, if the two-conductor problem is solved, the one-conductor ground-plane problem is also solved.

The exact electric field coupling (capacitive coupling) between two conductors over a ground plane or between pairs of conductors is a function of geometry, wire size, loading, terminating impedances, and spacings. To place the problem into perspective it helps to have a few milestones. When one geometry is known, simple extrapolations can be used to treat other problems.

Consider two parallel conductors over a ground plane or two parallel wire pairs as in Figure 4.4. If the conductors are #22 AWG and the circuit impedances are 100 ohms, then at 1 MHz the voltage coupling per meter of length is -62 dB. This coupling is roughly proportional to wire length, frequency, impedance, and inversely proportional to horizontal spacing. If the value of h is 0.1 cm, the coupling would be -81 dB.

Magnetic cross coupling (current to current) can often exceed capacitive coupling. The coupling per meter for the circuits shown in Figure 4.4 is -47 dB. The coupling is roughly proportional to wire length and frequency and is inversely proportional to impedance and the square of horizontal spacing. If the value of h in Figure 4.4 is 0.1 cm, the coupling would be -79 dB.

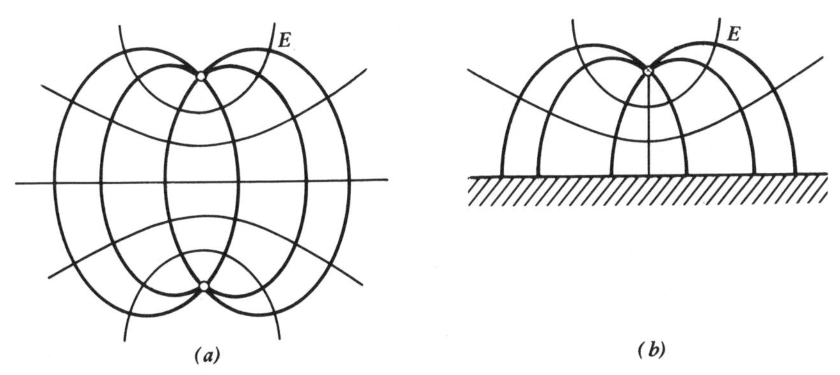

FIGURE 4.3 *(a)* The electric field between two conductors and *(b)* a conductor over a ground plane.

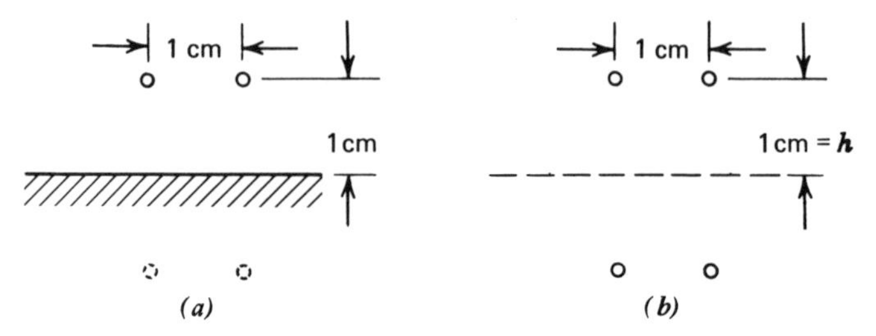

FIGURE 4.4 Coupling geometries.

4.4.1 Power Line Coupling

It is standard practice to keep reasonable separation between power or control lines and signal lines. It is worth calculating the magnetic power coupling for a 100 m run of 100 A source located 1 m away from untwisted signal lines. The magnetic field from the power line has a maximum value of

$$H = \frac{0.1Is}{D^2} \tag{4.4}$$

where H = magnetic field in amperes per centimeter
s = spacing between power conductors in centimeters
D = distance from power line in centimeters
I = current in amperes

The flux ϕ picked up in 1 m^2 is 0.01 weber. The voltage calculated by Lenz's law is:

$$V = 10^{-4} \frac{d\phi}{dt} \tag{4.5}$$

At 60 Hz

$$V = 360 \ \mu\text{V peak}$$

It is apparent from the above analysis that high-frequency current transitions on the power line can easily cross couple millivolts of signal. Good separation of power lines from signal lines is thus a necessity. If the power bundle is slowly twisted and if the signal pairs are twisted, cross coupling can be significantly reduced.

4.5 LIGHTNING

Lightning strikes are significant pulses of energy entering the earth plane. It is not uncommon for current pulses to exceed 100,000 A. The peak power can often exceed 10^{13} watts. More impressive is the 400 megajoules that can be dissipated in a single strike. Integrated circuits can absorb about 10^{-5} joules before damage results. A typical strike lasts about 20–30 μs even though the time appears longer.

The frequency spectrum of a pulse can be deduced by considering its rise and fall times. The current in a pulse rises from 10 to 90% of final value in time $\tau = 0.5\mu$s and falls to half-value in time $\tau_2 = 25\mu$s. The spectral response of this trapezoidal waveform can be found in the literature. The frequency $1/\pi\tau_1$ associated with the peak current can be used to approximate the voltage drop in any reactive circuit.

The voltage gradients along the ground plane in the vicinity of a lightning strike can be quite damaging. For example, at 100 m from a direct hit the gradient can be 1000 V/m. At 1 km from a direct hit the gradient can be 15 V/m. These gradients are thus impressed between two pieces of equipment that are separately grounded. The signals that result from this voltage gradient are impressed as common-mode signals on any interconnecting lines. To insure that these potential differences will not damage the hardware, voltage limiters must be applied to all conductors.

The problem of protecting hardware against surge voltages in not as simple as it might first appear. The difficulty arises for two reasons:

1. It takes a finite time for a suppressor to fire.
2. If several suppressors are required, they may not fire at the same moment in time.

The first problem may allow a high surge voltage to propagate beyond the supressor before the voltage is shunted. This certainly shortens the pulse length, but damage may still result from the shorter pulse. The second problem results in a temporary shift in circuit potentials that can cause damage. This phenomenon is illustrated in Figure 4.5. If surge arrestor C fires first, then $E_1 - E_2$ suddenly becomes 100 V. Until zener A conducts, the equip-

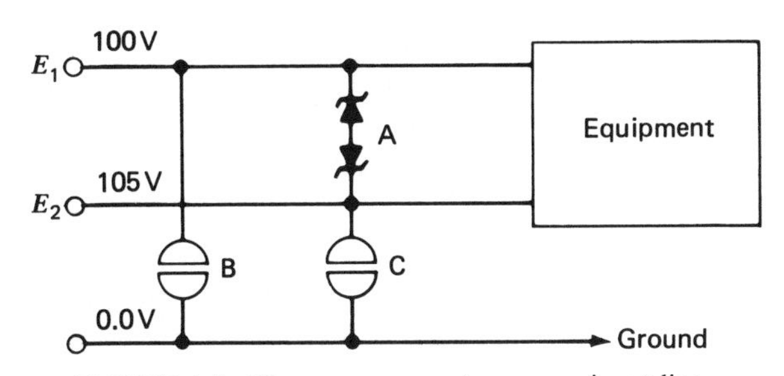

FIGURE 4.5 Three surge arrestors on an input line.

ment sees a difference potential of 100 V. Of course, when surge arrestor B fires, zener A is no longer involved.

The key to successfully protecting equipment against surges is to accept the surge arrestor delays and filter any subsequent pulses that get around the arrestors. Thus placing arrestors on incoming lines may not stop damage from occurring.

The leads that connect to an arrestor are inductive and the geometry of the connection may reduce the effectivity of the arrestor. Thus the physical location of the arrestors and their circuit geometry is critical. Arrestors must ground to the framework of the equipment being protected and not to a separate ground or "green" wire.

A variety of protection devices are available on the market. Metal oxide varistors (General Electric) have a nonlinear voltage-current curve that can be used to protect power equipment. Another product is a back-to-back PN semiconductor junction (General Semiconductor Trans Zorb®). These devices are extremely fast and can dissipate significant amounts of peak power.

Arrestors should be placed at building or equipment entrances and not inside equipment. Once inside the equipment, lines may arc across to other conductors before ever being suppressed. At the point of suppression the resulting short circuit causes a reflected pulse that propagates back toward the source. This reflected pulse is of opposite polarity and should also be considered a possible threat.

4.6 TELEPHONE LINES

Open telephone lines are used occassionally to handle analog signals. The lack of proper shielding precludes their use for low-level wide-band signals. If adequate input filtering is provided, low-level narrow-band signals can be handled.

High-level low-impedance sources can be used to drive long telephone lines. Because line resistances can reach thousands of ohms, a ground loop cannot be tolerated. The best approach to avoid this problem is to provide a differential buffer at the receive end of the cable. This solution is shown in Figure 4.6. If the resistors are matched to 0.1%, the ground potential difference will be attenuated by 60 dB. The current that flows between grounds is limited by the value of R. Typical values might be 100 kohms.

Telephone lines are usually "loaded" to maintain a 3 kHz frequency response. This "loading" consists of balanced inductors placed at fixed intervals along a telephone wire pair. The lines themselves are also balanced and terminated in 600 ohms. These techniques are difficult to apply in dc instrumentation. For these reasons the frequency response of untreated tele-

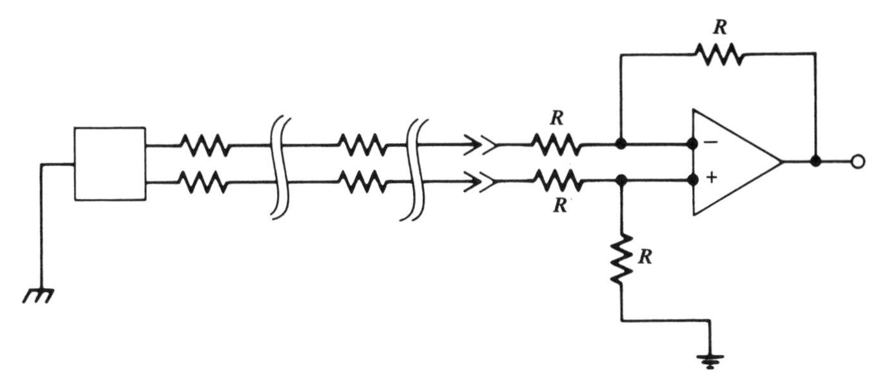

FIGURE 4.6 A "receive" differential isolator.

phone lines is apt to be poor. Some frequency compensation is possible in amplification that follows the differential isolator shown in Figure 4.6. Of course, if the cable run is short, the frequency response may be adequate.

In some telephone systems the earth is used as a conductor for ringing subscriber bells. Here the ring voltage is impressed between one side of the line and the earth. Ring voltages may vary in frequency and in amplitude and are often superimposed on the central office 48 V battery. This ringing current is just one more noise source in the ground plane.

4.7 SHIELDED CONDUCTORS

Shielding as applied to input signal lines is discussed in Chapter 3. Here a two-conductor shielded cable is recommended for input signals. Strain-gage bridges require up to ten conductors in a single group. The shield is an insulated conductor that surrounds the signal leads. Groups of shielded signal conductors can be supplied in one bundle, which can also have an overall insulated shield.

There is a frequent misconception regarding coaxial cable and shielded wire. The coaxial sheath does act to shield the inner conductor, but the shield is also a signal conductor. A coaxial connection requires that the outer conductor be connected at both ends or the circuit is broken and is certainly not coaxial. Connecting the shield at both ends often forms a ground loop that can allow differential coupling (see Section 4.3.2). If the sheath carries a ground connection forward without forming a ground loop, then the sheath serves as a shield, a signal conductor, and a coaxial connection.

The sheath of a coaxial cable conducts any externally induced currents to ground. As discussed earlier this unwanted current can cross-couple and become a differential-mode voltage. This coupling is a function of cable type, cable length, and the frequency of the interference. Obviously, signal-to-noise ratios must also be a consideration. Because of this coupling, long, open coaxial runs are not recommended for low-level instrumentation.

Long cable runs must often handle wide-band signals. Twisted shielded pairs are often not suitable for this type of transmission. Under these conditions a coaxial cable can be used, but with a second shield to drain external currents. This configuration is shown in Figure 4.7. This added shield may only be necessary when low-level signals are processed.

High-frequency transmission over a cable requires a constant characteristic impedance. This can only be achieved by controlling the geometry of the cable. Concentricity, uniformity of dielectric, and wall finish are all important considerations. For long signal lines, the frequency response can only be controlled by terminating the cable in its characteristic impedance.

The luxury of terminating each cable in its characteristic impedance is not available in most instrumentation processes. When a terminating resistor is used, the signal loss due to cable resistance can be significant. This loss can be accommodated through calibration at the time the signal is processed. Note that this loss is temperature sensitive because the cable is made of copper. Transducer impedances rarely match the characteristic impedance of the cable they drive. If the cable is ideal, the energy entering the cable from a mismatch would not cause a problem assuming the receive end is properly terminated. Without this proper termination, energy returned from the far end will rereflect at the source mismatch.

When the frequency response of a cable run is important, a frequency response measurement is recommended. Adjustments can then be made to the termination to optimize the response. Frequently, a series *RC* termination can be very effective. A square-wave test is a good way to find values for *R* and *C*. The test circuit is shown in Figure 4.8.

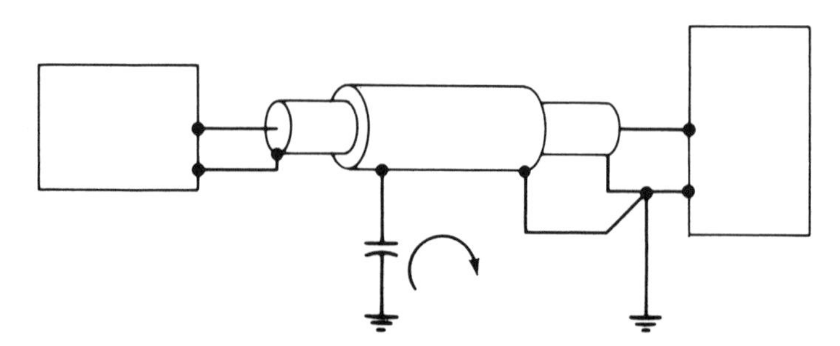

FIGURE 4.7 A second shield to drain unwanted current in a coaxial connection.

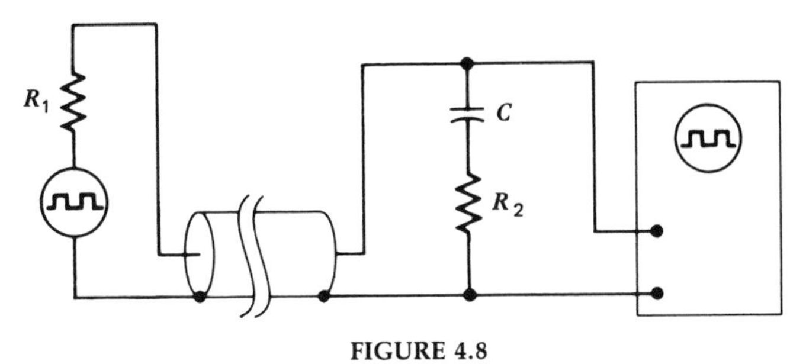

FIGURE 4.8

The first step should establish the low-frequency response of the cable. A square-wave frequency of 100 Hz can be used to measure the output voltage after all transients have died out. This waveform is shown in Figure 4.9. The square-wave frequency should then be increased until the leading edge definitely shows transient pheonomena as in Figure 4.9*b*. The values of R_2 and C can then be adjusted until a pattern similar to Figure 4.9*c* is obtained. The final values in (*c*) should equal the final values in (*a*).

4.8 NOISE AND VIBRATION

In high sound pressure ambients, unwanted signals are often generated by mechanical vibration. This can be the result of cable flexing or conductors moving in a magnetic field. If this phenomenon is expected, then some sort of signal-to-noise ratio testing should be performed.

In tests involving explosions or shock waves, the compression of the ground plane can cause transient common-mode voltages. The nature of this signal can be determined by dedicating an instrument to common-mode monitoring. If the levels encountered are within the limits of the instrumentation, the data are validated.

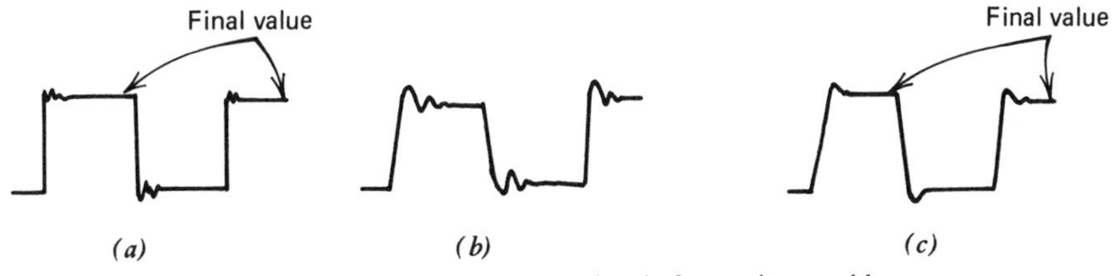

FIGURE 4.9 Square-wave signals for testing a cable.

4.9 CABLE RUNS—RACEWAYS AND TRAYS

Shielded signal lines are relatively free from differential electromagnetic interference. When these lines are inside raceways or trays, they are further protected by these added metal housings. Frequently, raceways or trays are open on the top or the covers are not tight fitting. This indicates that interference can be picked up from nearby fluorescents or from transient phenomena on parallel power lines. This pickup, however, may only be a problem in high ambient situations.

For a raceway to be effective at rf, all seams and covers must be electrically tight. This can be accomplished by welding or rfi (radio-frequency interference) gasketing all seams.

CHAPTER FIVE

INSTRUMENT ENVIRONMENT AND PLANT

"As is the workman so is the work."

Anyone who has opened the rear doors to a rack of hardware has seen the maze of cables that comes up out of the floor or down from the ceiling. Another maze involves the cables laced together under the subfloor or the cables brought into a distribution panel. Not to be neglected is the experiment that has grown like Topsy with wires crisscrossing in every direction.

Order cannot be made of a final installation by visual inspection. One path to a meaningful understanding is through schematic representation. This also fails, however, if the location of shield ties, ground point, filters, and so on, is not fully described. Furthermore, schematics rarely describe the separation of cables, relative proximities, and wire lengths. A diagram that shows these is not a standard schematic; it is attention to these details that makes for a good system.

5.1 POWER ENTRANCES

Building power and the service transformer are often located at one side of a building. The distance between hardware and this transformer can be from 10 ft to 3000 ft. Often the power lines, neutral and safety ground, are brought into the instrumentation room over this same distance. The inductances inherent in these lengths would imply that these connections are ineffective at rf.

In the United States, the code requires that all electrical frameworks that can be touched by humans must be safety grounded. This grounding cannot be via the neutral power connection. This ground or "green wire" must be a dedicated conductor valid at 60 Hz and must be earthed per local code. In a typical installation many devices are connected to this safety ground structure in a random sequence. This means that parasitic currents that flow from filters, transformers, entering cables, external pickup, and so on, flow in an ill-defined ground grid. The potential difference between any two pieces of equipment will be affected by this grid geometry and by the magnitude of any currents that flow.

Consider the case where a ground or green wire services both an instrumentation facility and some additional equipment (typewriters or lights or fans). In this situation ground potential differences appear as power line common-mode signals (see Figure 5.1). Because of the inductance in each ground connection, this common-mode signal is apt to be high frequency in nature.

In Figure 5.1, $I_1 + I_2 + I_3$ flowing in Z_2 causes a power line common-mode voltage for equipment II, whereas $I_1 + I_2$ flowing in $Z_1 + Z_2$ causes a common-mode voltage for equipment I. In the hardware itself this power line common-mode signal can have various effects that depend on hardware class, filter location, wiring geometry, internal transformers, and so on. Another problem arises when voltage E_1 is a common-mode voltage between equipments I and II.

Many large pieces of hardware have built-in line filters that bypass high-frequency currents to ground. This current flows directly to the green wire without entering the hardware proper. In Figure 5.1 the filters may improve the local performance, but the currents flowing in Z_1 and Z_2 may not be

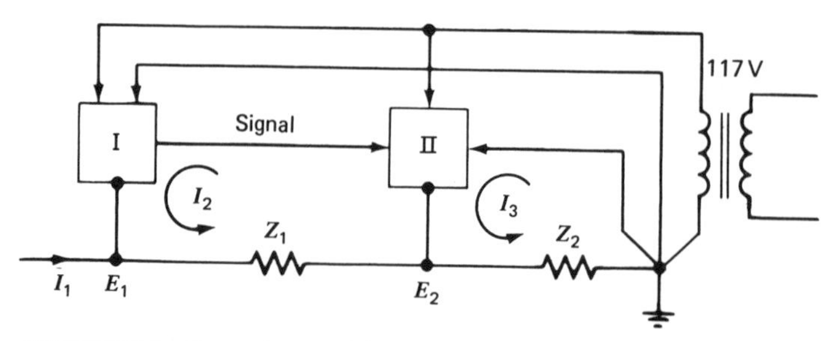

FIGURE 5.1 Ground potential differences as common-mode signals.

significantly reduced. This means that the common-mode voltage between devices I and II is not removed through the use of power line filters.

One approach often tried is to take a ground from each piece of hardware back to a mecca on a separate safety wire. This would increase the value of Z_1 and it might also increase the common-mode level between devices I and II.

One viable way to resolve this problem is to reduce the value of Z_1 (Figure 5.1). This can be done by using a ground plane, shortening the grounding path, or using a better ground connection. A second viable method is to filter the signal at device II. Finally, an optical signal link could be used and this would be insensitive to common-mode potentials.

The problems discussed above are critical in digital hardware or where high-frequency signals are being processed. If high-frequency common-mode signals overload any low-level instrumentation, the problem can be equally severe.

5.1.1 Local Distribution Transformer

An effective way of powering an instrumentation facility is to locate a separate power transformer near the system hardware. This allows using short power lines, a separate earth ground, and a short safety conductor. This method is effective only if noncritical devices that are powered through separate power lines carrying their own safety conductors. This method is shown in Figure 5.2. This separate transformer (B) can be shielded and the secondary filtered to ground. This further reduces the flow of current in the separate safety conductor. Transformers with this shielding and associated filters are commercially available. Note that transformer (B) need not be powered from the mains directly. The secondary of transformer (A) may be at a lower voltage that is more suitable to building distribution.

Transformer (B) is effective even without shields. It isolates one set of power connections from another. In this sense it is the proverbial isolation transformer.

5.2 THE GROUND PLANE

Buildings constructed with steel and concrete usually have reenforcing bars buried in the concrete floor. These bars and other associated conduits make up a pseudo ground plane under equipment. The electrical connection between bars is usually uncertain because welded joints are rarely specified. Because this ground is poorly defined, a special ground plane is often con-

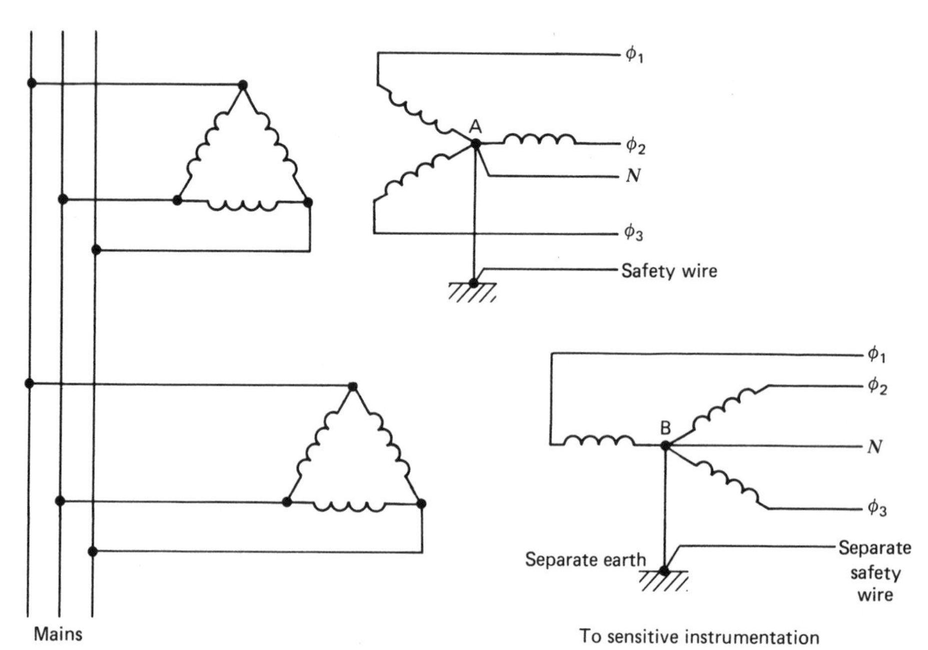

FIGURE 5.2 A separate power transformer for instrumentation facility.

structed above the floor and under computers and their related peripherals. To be effective this ground plane should be safety earthed at one point and not connected to building steel or other building conduit. This procedure limits the ground plane current to local phenomena. Note that air ducts, raceways, and power conduits are all potential ground current paths.

Power line filters should return unwanted current to the ground plane via a very short path. If the ground-connecting lead is long, it is inductive and the filter is ineffective. Filters should therefore be mounted near the ground plane or on the outside of equipment housings that are connected to the ground plane. It is important to note again that unfiltered signals should never be brought inside hardware to be filtered. When this is done, entering leads can radiate to all other circuitry and the filtering may be ineffective.

It is practical to isolate ground connections at rf by using inductors and/or ferrite beads. These connections can still serve as safety grounds at 60 Hz, but can impede the flow of high-frequency ground current. For lightning protection these inductances should be bypassed for high-voltage pulses. This circuitry is shown in Figure 5.3.

Ground planes can be built in many ways. A grid of copper strips buried in the floor or under floor tiles is satisfactory. A solid sheet of metal is an

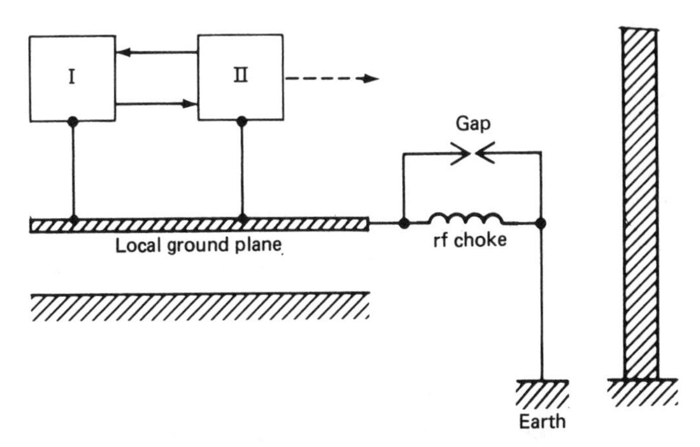

FIGURE 5.3 An isolated ground plane with lightning protection.

excellent solution. This sheet of metal need only be a few mils thick to be effective. It is important that all seams and joints be carefully soldered or welded. If the ground plane is to be practical, convenient connecting points need to be prearranged. Because heavy straps may be used, these connections must be mechanically strong.

5.3 DIGITAL INTERCONNECTIONS

Most facilities will have a mix of analog and digital hardware. Digital signals have high-frequency content that can disturb other sensitive hardware. In some installations digital and analog processes are necessarily intermixed. An example might be digital control of calibration or gain in an instrument amplifier.

The rise time for TTL logic is about 1.5 ns. This represents a frequency of $1/\pi\tau = 210$ MHz. This is a high enough frequency to easily radiate and cause slew-rate overload in analog instruments. It is desirable to avoid bringing digital signals into the analog environment except under well-controlled conditions.

It is necessary to keep analog grounds separated from digital grounds. The difficulty arises when logic currents flow in the analog commons. The high-frequency content together with lead inductance cause momentary overloads and subsequent recoveries; thus the noise that is detected may not look like the signal that caused the problem. Signal rectification is common and the net effect can be a dc shift or "demodulation."

If a digital device is to control a group of instruments and the instruments terminate at separate points, then any ohmic connection to each instrument causes ground loops. For this reason some form of isolation is a necessity for the digital control of instruments.

The logic that controls an instrument can often be limited in bandwidth. This filtering can be done at the logic driver or within each instrument. Bandwidth reduction does not remove the ground-loop problem, but it may eliminate unwanted coupling in the instrument proper.

5.4 MULTIPLE OUTPUT LOOPS

It is frequently desirable to connect the output of an instrument to several points. Connections might be to a tape recorder, an alarm panel, or to a computer. Each of these devices has its own ground. Even if these grounds were at the same potential the ground loops that result can still cause problems. This is illustrated in Figure 5.4. Note that the amplifier provides a ground connection between the tape recorder and the computer. When more equipment is involved the instruments provide a complex set of ground ties.

If load current is demanded by one device, some of the current can return via other cables. For short runs this is not a problem. Note that at high frequency the cable capacitance can be considered a significant load. Thus

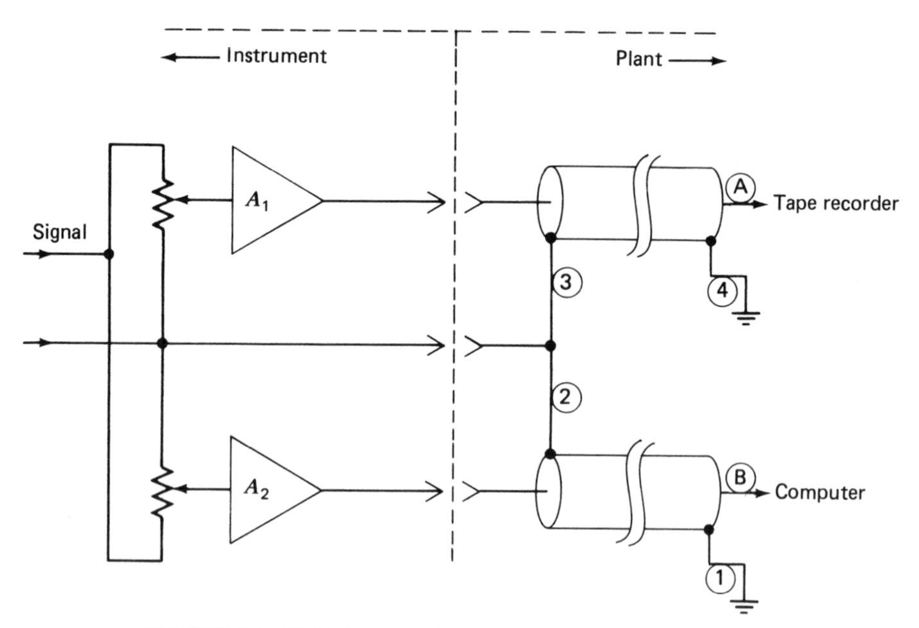

FIGURE 5.4 Multiple output loops. Note ground loop 1–2–3–4–1.

multiple return paths can cause some signal cross coupling, depending on cable length and frequency.

Instruments can be designed to avoid this ground loop. A separate power supply and an isolating amplifier can be used to drive one of the cables. This solution is shown in Figure 5.5. The ground-loop impedance here is 200 kohms.

Some instrumentation systems utilize a common power supply for up to 16 channels. Connections between the instrumentation and other grounded devices cause ground loops. When common supplies are used and there is circuit gain, systems can become unstable. Because this is a systems problem, this instability cannot be detected by testing the individual amplifiers.

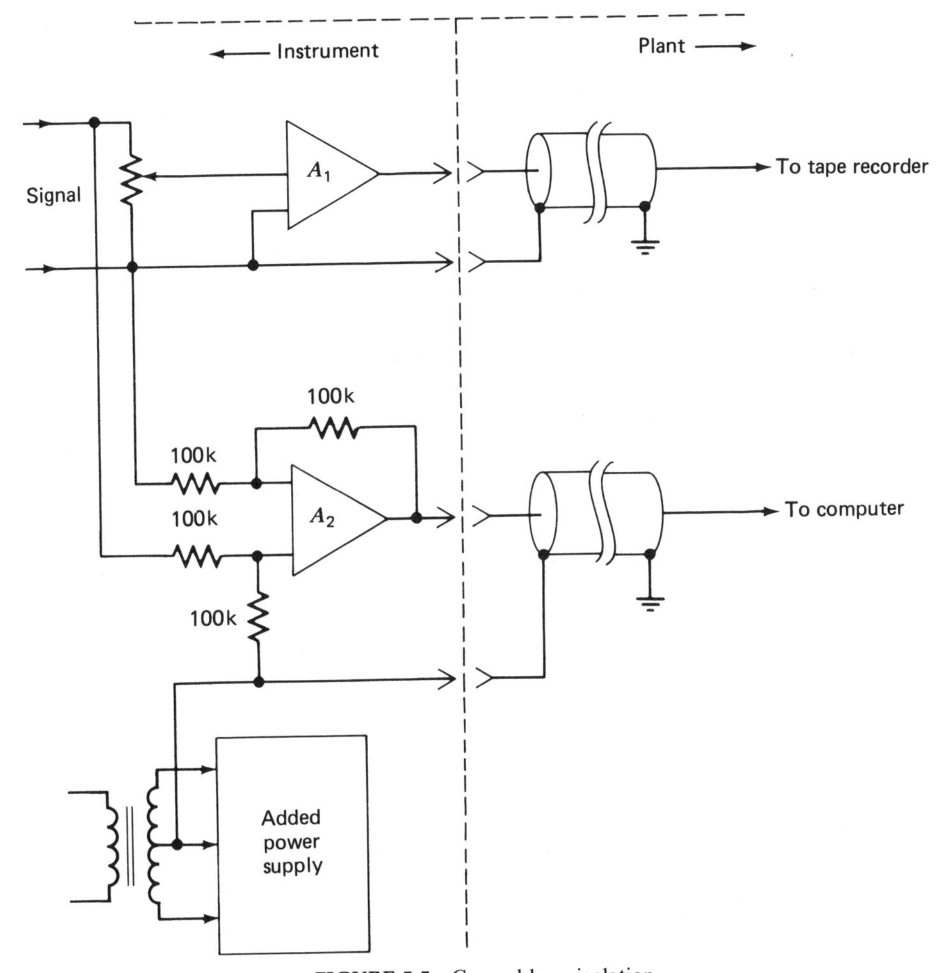

FIGURE 5.5 Ground loop isolation.

5.5 DISTRIBUTION PANELS

Facilities are usually built with flexibility in mind. This flexibility requires that cables be terminated at a distribution point so that transducer assignments and amplifier assignments can be varied according to test requirements. To avoid improper shield connections, the connector shells need to be insulated from their mounting surface or mounted on an insulated panel. If the shield is brought through on a separate pin and the connector shell is improperly grounded, a portion of the signal path will be open to contamination. For low-level signals this would be a source of noise pickup. The connector shell needs to be tied to the signal shield, not to an undefined ground.

Low-level signals require symmetry to avoid thermocouple effects. In a multipin connector, adjacent pins should be used for input signal connections. A small loop area also reduces the possibility of magnetic field pickup.

Low-level signals should interconnect on gold-plated contacts. This is the best protection against contact noise. The resistance of the connection is not as important as its stability with time. Patch cords are often used to interconnect these low-level signals.

5.6 RACKS AND RACK MODULES

Instrumentation facilities rely on rack cabinets to house their equipment. Racks are available with front and rear doors, top or bottom ventilation, extra depth, rfi shielding, power ducting, and so on. As discussed earlier, rfi shielding can be easily violated by unfiltered signal or power entry. Racks are selected based on the environment, building design, and expansion needs.

All racks must be safety grounded. Codes specify the wire size and the quality of the earth connection. It is preferable that each rack cabinet be grounded separately. If one rack is grounded through another, local code specifies the maximum spacing that is permitted. Safety grounds should be made through separate conductors and hardware. All paint or anodization must be removed before connections are made.

Hardware that mounts in a rack cabinet is grounded by that rack. In some instruments this forces the ground of the instrument to be tied to rack ground at this point. This situation occurs for most oscilloscopes, signal generators, and voltmeters. This ground can be avoided by using mounting insulators. A ground for the instrument must be provided by the system when the instrument floats from the rack.

Instrumentation housed in a rack module is usually designed to be totally isolated from the rack cabinet. The instrumentation housing (rack module) is grounded, but the instruments themselves are electrically floated. The metal housing around an individual instrument is usually tied to input guard or output common. All connections to this outer shell should be made by the system not by the rack cabinet.

Housing instruments in rack modules allow easy removal for service and calibration. This feature usually requires a double set of connectors—one set on the rear of the rack module and another set at the instrument/module interface. This may necessitate a large number of internal interconnecting cables with many of them shielded. If, however, the instruments are built with connectors on their own rear surface, a double set of connections can be avoided. When adequate service loops are provided, instruments can be removed without rear access.

5.6.1 Ventilation

An important consideration in any instrument facility is ventilation. Air for cooling can be taken from the room, the floor, or the ceiling. Cooling can occur by forced air or by convection. Manufacturers of equipment must guard against a worst-case condition unless there has been some prior arrangement.

The vertical stacking of hardware poses many problems. Some designs require a free flow of vertical air, whereas others rely on fans to pull air horizontally. Some units dissipate a great deal of power, whereas others are nondissipators but very sensitive to temperature gradients.

Fans in themselves pose a problem. If the air is properly prefiltered, dirt buildup will not be a serious problem. Individual air filters in instruments require regular maintenance if they are provided. Fans are also subject to failure and unless warning systems are present, damage can result.

Low-level instrumentation is particularly susceptible to thermal gradients and thus to air flow. Each connection along an input path is a potential thermocouple. If the leads are heated by hardware in the rack and cooled by air conditioning in an unsymmetrical manner, error signals will result. This is particularly true if the air conditioning is cycled.

5.7 ISOLATION TRANSFORMERS

A variety of isolation transformers are available on the market. These transformers are supplied with one, two, or three electrostatic shields. It is left to users to correctly connect these shields to their system.

Isolation transformers that power more than one device provide limited signal isolation. This problem is shown in Figure 5.6. Regardless of how the

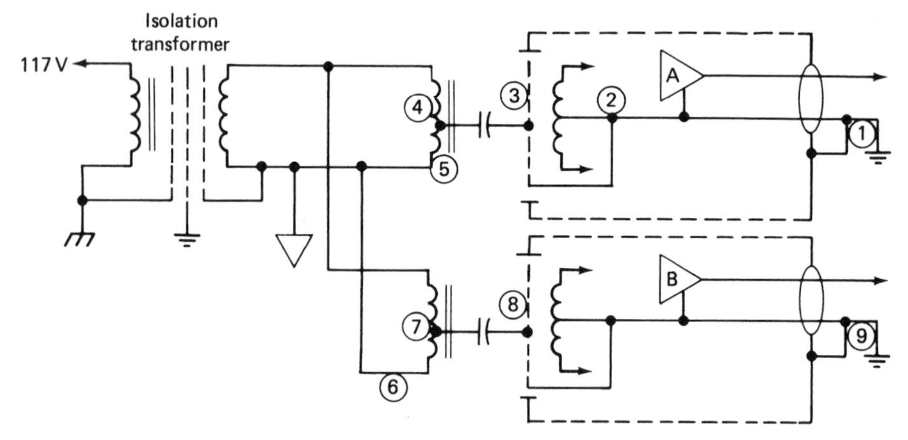

FIGURE 5.6 An isolation transformer problem.

shields are connected in the isolation transformer, currents will circulate between devices A and B in path **1–2–3–4–5–6–7–8–9**. This loop includes transformer voltages **4–5** and **6–7**. If devices must be isolated by adding a shielded transformer, each device needs its own transformer.

An isolation transformer can be used for reducing common-mode effects on power lines. (A common-mode signal is a common voltage between all power lines and the local ground.) Without proper shielding these common-mode signals can enter the system commons. These signals are often coupled into the power line/ground loop by external rf fields. The transformer connections for common-mode treatment are shown in Figure 5.7.

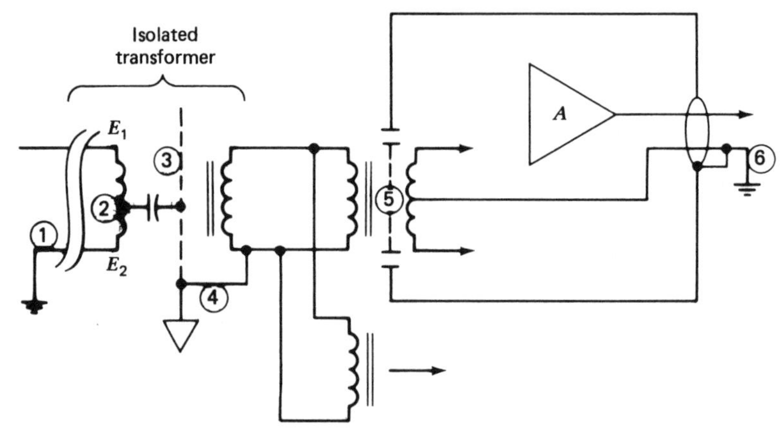

FIGURE 5.7 Rejecting power line common-mode signals.

Common-mode voltage E_1 forces current flow in loop **1–2–3–4–1**. Without an isolation transformer, coil **2** abuts shield **5** and common-mode voltage E_1 forces current to flow in loop **1–2–5–6–1**. This current flows in output common lead **5–6** and can cause trouble at high frequencies.

Power is delivered to a transformer in a differential mode. This voltage difference causes flux changes that couple to the secondary coils as voltage (transformer action). Shields cannot be used to reduce differential-mode noise any more than they can be used to reduce power. If differential-mode noise signals are present, a transformer will provide some filtering because of its inherent leakage inductance and winding capacitance. If this built-in filtering is inadequate, a passive line filter may be necessary. The shields can, however, be used to control noise current flow.

Differential-mode noise results from step loads, arcing on inductive interrupts, pulse loads, and so on. This noise is usually filtered out by internal power supply electrolytics. The noise currents do, however, enter the instrument or device and therefore can radiate to other parts of the circuit. This is the main argument for line filters. A line filter can remove differential power line noise so that it does not enter the devices in the first place.

The isolation transformer in Figure 5.7 has only one shield. One must then ask what the other shields are used for. If an isolation transformer has three shields, where should they be tied? One shield should be connected to the "green wire" as shown in Figure 5.7. The remaining shields can be used to control some aspects of differential mode current flow. Figure 5.8 shows this technique.

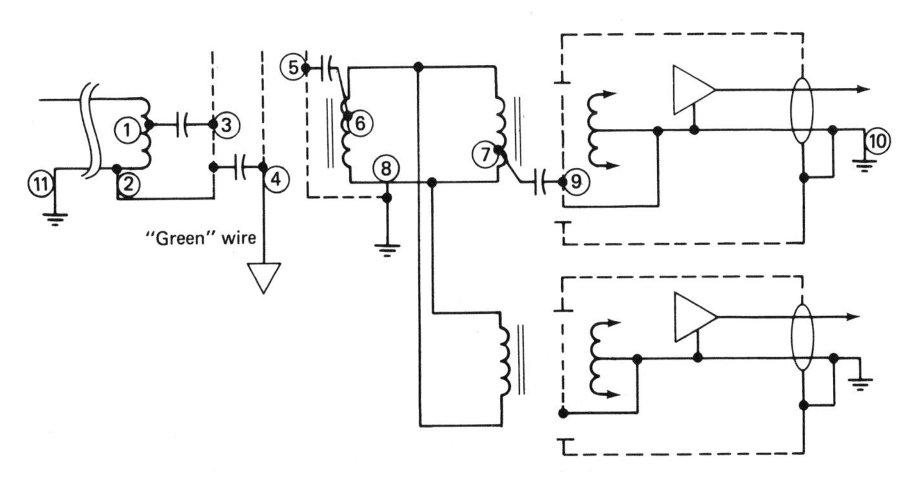

FIGURE 5.8 Connections for three shields in an isolation transformer.

Differential-mode noise voltage V_{12} circulates current in loop **1–3–2–1** through capacitance C_{13}. This current would flow in capacitance C_{14} if shield **3** were removed. This current would thus use the green wire for its return path to **11**. In a properly designed system this would not be a problem. In some systems green-wire current causes common-mode signals between devices (see Section 5.1).

A similar argument can be used for differential-mode signal V_{68}. Current here flows through capacitance C_{56} in loop **5–6–8–5**. If shield **5** were removed, this current would circulate in the green wire again. The final insult, however, is that differential voltage V_{78} flows in loop **7–9–10–8–7** and this flows in signal conductor **9–10**. This current cannot be controlled if the isolation transformer is shared by more than one device. Note that multishielded isolation transformers can reduce green wire current but not current flow in signal commons.

If the differential-mode noise current in conductor **9–11** is to be controlled, a dedicated isolation transformer must be used (see Figure 5.9). Here noise voltages on coil **1** circulate current in loop **1–2–8–1**. Noise voltages on coil **4** circulate current in loop **4–3–9–4**. Noise voltages in coil **5** circulate current in loop **5–6–9–5**. Finally, noise on coil **7** circulates current in loop **7–6–9–7**. Note that common-mode voltages **1** and **8** circulate current in the green wire.

The isolation transformer in effect substitutes a multishielded transformer for the transformer within the device. Shield **3** is connected to the device common and thus the isolation transformer cannot be shared with other devices.

When connecting a multishielded transformer, care must be taken not to violate any of the shields. In Figure 5.9 the interconnecting lines must be shielded by conductor **9** and not be the green wire. Similarly, lines **1** and **8**

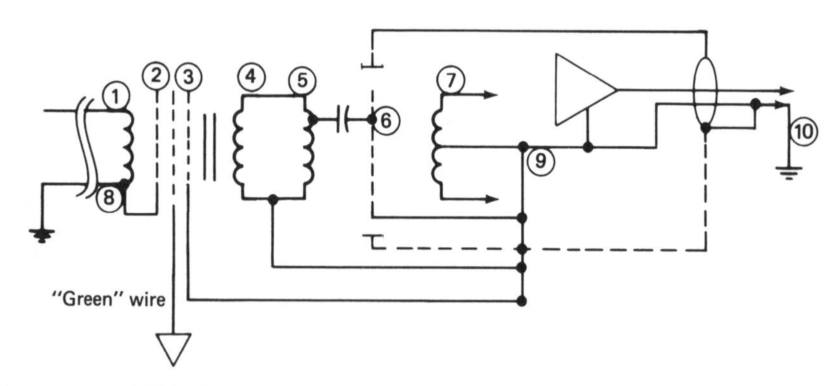

FIGURE 5.9 An isolation transformer applied to one device.

must be shielded by potential **3**. This means that if conduit or shielding is used, it cannot be indiscriminately grounded.

If the green wire is inductive, common-mode voltages on lines **1** and **8** can circulate current in conductor **8–9–10**. In effect, the middle shield becomes ineffective. Thus the isolation transformer should be mounted at or near the rack cabinet being serviced.

CHAPTER SIX

SIGNAL CONDITIONING

"Gains when poorly gotten will be sure to turn rotten."

The following chapters are intended to familiarize the reader with many facets of instrumentation design. This chapter treats input signal conditioning and later chapters discuss the amplifier design proper.

Several facets of signal conditioning have already been treated. For example, Section 3.7 discussed input filters and RTI offset and Section 3.4.3 discussed preamplifiers. This indicates again that instrumentation is a systems problem and that not all answers are found within the instrument amplifier.

6.1 INPUT GUARD SHIELD

The input signal shield is often called a guard shield because it presumably "guards" the input. The need to maintain its integrity is covered in Section 1.6. The shield should "guard" the input leads and the input circuitry. Ideally the input shield should be connected to the signal common at the transducer and make no further connections at the instrument. This ideal treatment is often violated because of other considerations.

The input shield assumes potentials at rf, depending on systems geometry and the presence of external fields. These signals are usually out of band both for common-mode rejection and signal processes and can cause overload and subsequent rectification (demodulation) within the instrument. To reduce this problem the shield is usually bypassed to the output common. This bypass does not impact in-band processes, but it does invite shield currents to flow in the output common. This circuitry is shown in Figure 6.1. The shield current that flows in R and C causes a potential gradient along the shield. This gradient reduces the effectivity of the shield at high frequencies. To avoid this situation, a separate shield can be used (see Section 4.7).

The RC shield bypass shown in Figure 6.1 could go to the output common through an output shield. This would keep common-mode current from flowing in the output common lead. However, if this lead were of any significant length, this bypass would not function. For this reason it is preferable to connect the RC circuit to the output common in the instrument.

Many circuits need to be at the input guard potential. These circuits include power supplies, relay coils, transformer shields, and outer wraps. Because the guard shield can drive a series 0.01 μF and 100 ohm load, it could also drive these secondary capacitances in the instrument. For several reasons, however, this is usually not done. Rather, a separate circuit is usually used to derive the common-mode signal and this circuit drives these capacitances. Note that the drive circuit need not be at the same dc potential as the input guard shield. It is only important that the dynamic gain be unity. A typical circuit is shown in Figure 6.2. The emitters of the input differential stages are summed through equal 100 kohm resistors to derive the average input signal or common-mode signal. The FET output is called the "driven guard."

A derived common-mode signal is necessary because users often misapply input shields. When the input shield is not correctly tied to source ground, the shield no longer represents the average input or common-mode signal. The only way the designer can force a measure of this common-mode level is to sum the two input signals as in the circuit of Figure 6.2.

The driven guard provides current to parasitic capacitances that would otherwise have to be supplied by sensitive circuits. This idea is shown in Figure 6.3. Here shield surface Ⓐ is tied to the driven guard shield. The

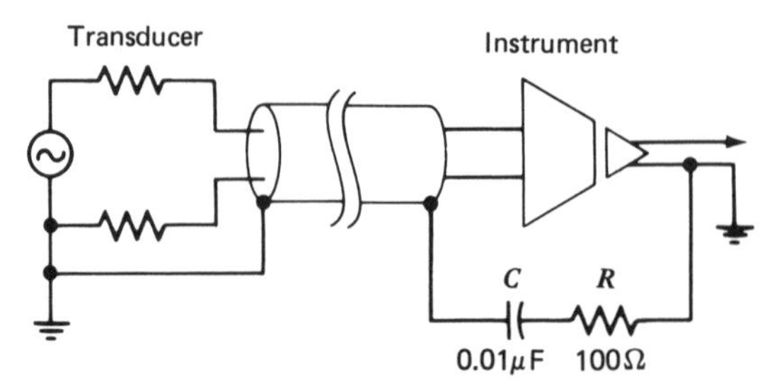

FIGURE 6.1 Input shield bypass.

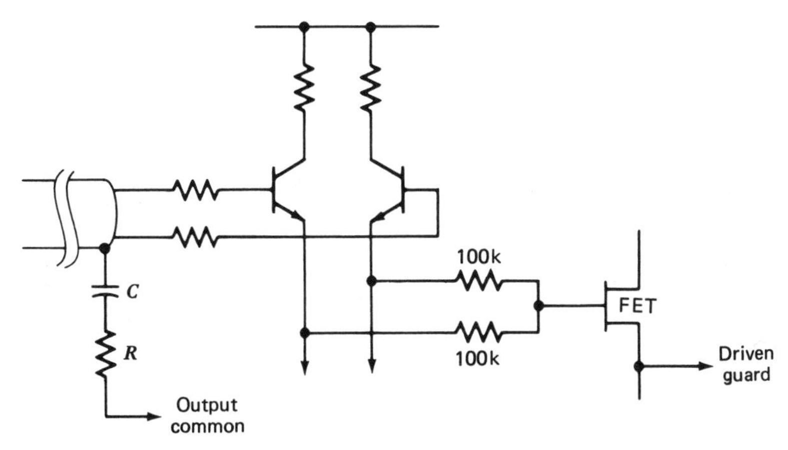

FIGURE 6.2 A "driven guard" circuit.

current in C_2 is supplied by the FET and the input lead need only supply signal current to capacitance C_1. The input lead does not supply current to capacitance C_2, which would degrade the common-mode rejection ratio. The regular input shield also performs this same function, assuming it is properly connected at the source.

6.2 INPUT FILTERING AND INPUT CIRCUITRY

Input leads frequently bring differential and common-mode signals to the instrument or device that are are out of band. These signals may exceed the slew-rate capabilities of the device and cause a form of signal rectification. This rectified error signal cannot be separated from normal signal once it is generated. To avoid this difficulty some rf filtering is usually provided at the

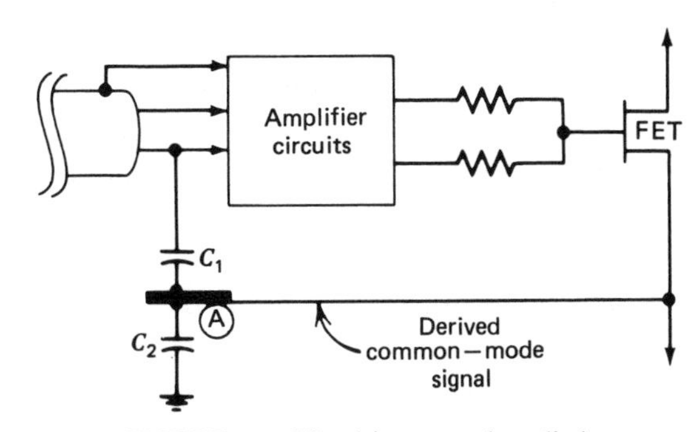

FIGURE 6.3 The driven guard applied.

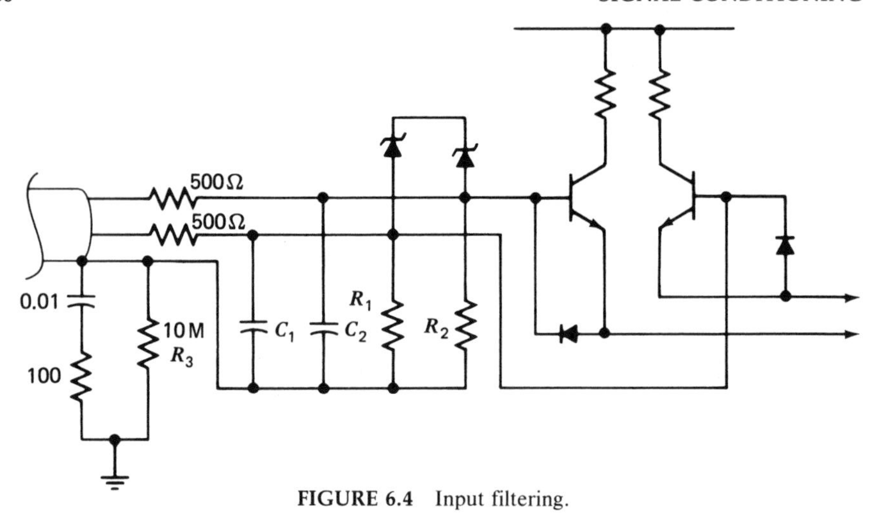

FIGURE 6.4 Input filtering.

input. This circuitry is shown in Figure 6.4. Capacitors C_1 and C_2 form a time constant with the 500 ohm resistors and any source impedance. This time constant is set to just above -3dB bandwidth of the device. Typical values of C_1 and C_2 are 100–500 pF.

Resistors R_1 and R_2 provide a dc return path for the input stage. This return path is through resistor R_3 back to the output common. This return path is required for applications where the input signal is not grounded.

Input stages are vulnerable to overscale signals. Zener diode clamps can be used to limit differential signals and back-to-back diodes can be used to limit base-emitter potentials. These clamps are shown in Figure 6.4. A typical zener diode voltage is 12 V. When the first active element is an integrated circuit, some of these clamping arrangements may be a part of the device.

6.3 THE INPUT ACTIVE ELEMENT

In any feedback system, performance is limited by the quality of the input or sensing circuit. Any error or noise developed in the system is reduced by the feedback factor and the gain preceding the point of error. Obviously, there is no gain ahead of the input device.

The input stage in most instrument-quality amplifiers is a matched pair of transistors made on a single dice. These pairs are often aged and hand-selected for their matched qualities. Some devices are laser trimmed. Of particular importance are differential offset, β, and input current. These parameters can be matched at one or more collector currents and voltages.

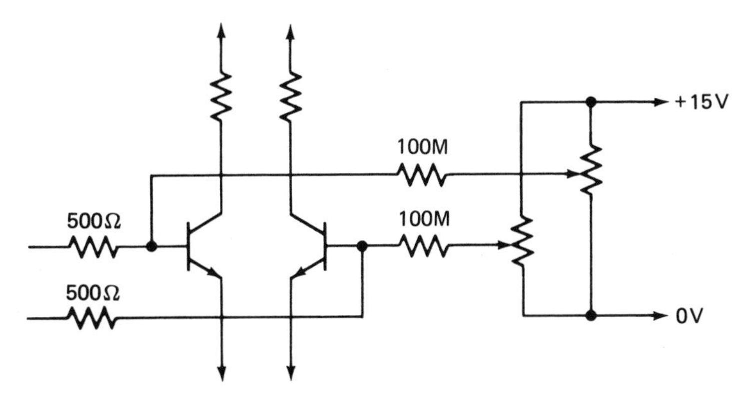

FIGURE 6.5 Base current cancellation.

Matching is critical for low dc drift and a low-temperature coefficient of drift.

A discrete balanced and matched pair of transistors has several advantages over an integrated circuit.

1. Collector current can be independently controlled. This is key to bandwidth control.
2. The collector circuit can accommodate a gain shelf at high frequencies. This open-loop response shaping can optimize frequency response.
3. The input collector resistors can be wire wound for added stability.
4. The noise figure is usually lower than that given for integrated circuits.

A typical β in an input stage is 100. If the collector current is 60μA, the base current is then 600 nA. This current flowing in a 1000 ohm source unbalance would cause an offset of 0.6 mV. To remove this error, the base current for the input stages must be supplied from an internal power source. This circuitry can reduce the source current to a few nanoamperes. In order to be absolutely correct, the base current has a temperature coefficient and the cancellation current should be temperature compensated. To avoid input loading, base current can be supplied through 100 M resistors. This circuitry is shown in Figure 6.5

6.4 DIRECT-IN SUBSTITUTION

A relay can be used to connect the amplifer input to a separate set of input connections. This second set of connections can be common to a group of instruments and can be used to calibrate the instrument proper.

The relay should switch all three input leads because a new guard potential needs to be sensed. If the secondary input is single ended, then the guard shield connection should be made to the source common. If the secondary input is wired as a two-conductor shielded cable, then the shield can be connected to the signal common at the signal source. This circuit arrangement is shown in Figure 6.6.

The relay used for direct-in substitution must not add to the amplifier drift. This drift can result from thermocouple effects along the input path. Each solder joint and relay contact are possible sources of trouble. These input junctions must be paired and kept away from temperature gradients and they must be kept symmetrical. The relay coil is one source of heat and thus thermal gradients. If the relay is used on a momentary basis, its heating may be insignificant. Reed relays use ferromagnetic materials and are therefore more susceptible to thermal problems.

The control leads and the relay coil itself are sources of signal contamination. These leads are connected to circuits that are not at the input ground

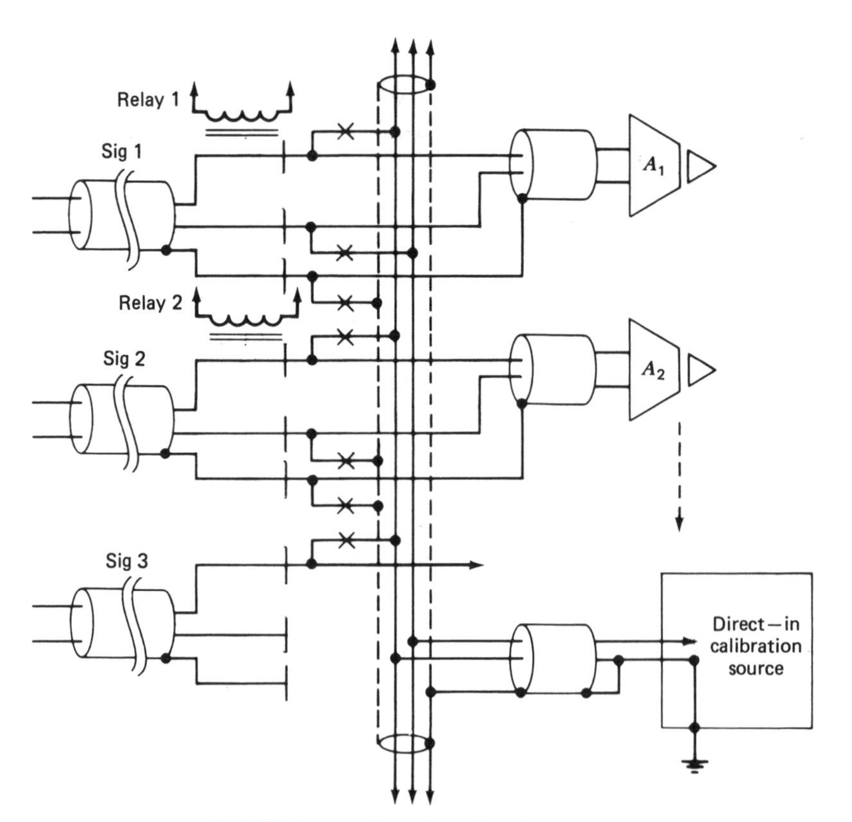

FIGURE 6.6 A direct-in calibration source.

potential. As indicated earlier, the leakage capacitance between the input leads and output ground must be held to just a few picofarads if the common-mode rejection ratio is to be kept high.

Several things can be done to reduce the input leakage capacitance caused by a relay. For example, the power for the relay can be derived from a source that is driven by the input common-mode potential. This in turn requires some form of isolation between the logic ground and the common-mode ground. Another less effective method is to tie the relay frame (if available) to the input common-mode potential.

The direct-in signal appears on the open relay contacts. Even if the signal is removed, these contacts are at a ground potential that can impact common-mode rejection ratios. For most contact geometries this leakage capacitance is negligible. It is possible to use guarding contacts in the relay so that a guard shield is always placed between each input lead and the direct-in signal. This, of course, increases the cost and size of the relay. The guarded contacts and the driven relay power supply are shown in Figure 6.7. This extra protection can easily be violated if the direct-in leads are not routed correctly.

A direct-in or calibrate signal can be routed to many instruments on a common bus. If the gain of an instrument is 1000, the calibrate signal must be below 10mV. Because instrument gains do vary, the calibrate signal must cover a wide dynamic range to be useful for all instruments.

It is often desirable to transmit only high-level signals for calibration. This technique requires direct-in attenuation within each instrument. With these attenuators ganged to gain selection, the calibration signal can be bused at high level for all instruments.

6.4.1 Solid-State Switching

Solid-state devices can be used to provide direct-in signal switching, but several problem areas must be considered. The signal potentials that are to be switched should not exceed the hold-off voltage of the switch. This places a definite limit on the common-mode potentials that can be accommodated. Most instrumentation common-mode levels are specified at 50 V with a no-damage provision of 300 V for short periods of time. Zener clamps can be used to protect the switches.

The ratio of "on" resistance to "off" resistance is much lower in a solid-state switch than it is in a relay contact. Typical on resistances values vary from 20 to 500 ohms, whereas off resistances are on the order of 10^8 ohms. The input impedance to an instrumentation amplifier is usually about 10^9 ohms. This means that the on resistance of a solid-state switch in series with an amplifier input impedance presents no problem. The guard shield switch

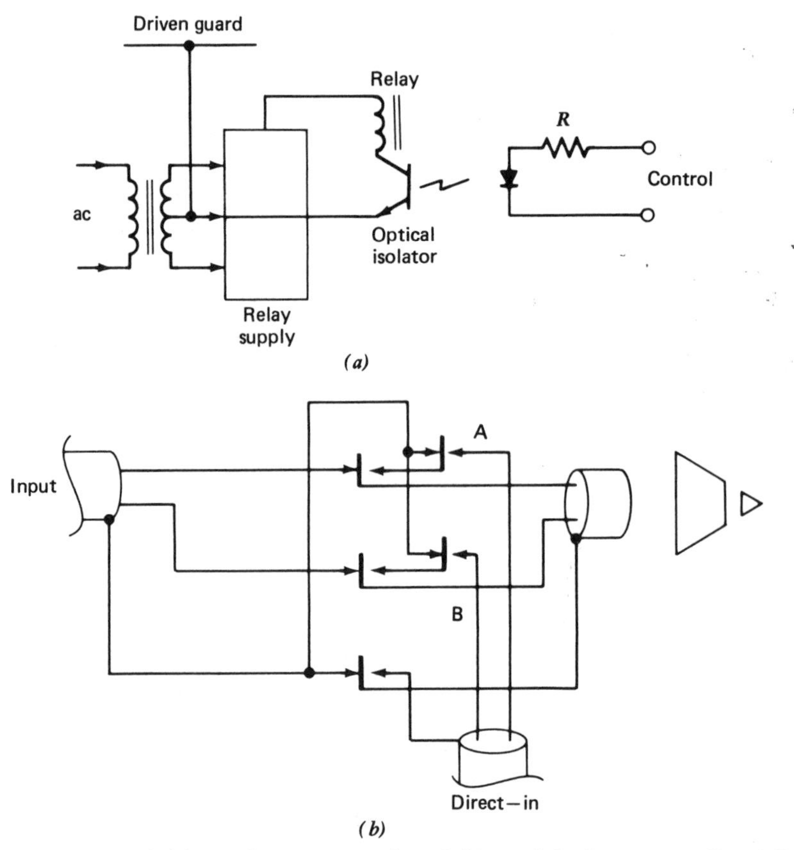

FIGURE 6.7 *(a)* A driven relay power supply and *(b)* guarded relay contacts Ⓐ and Ⓑ.

is adequate as long as the guard shield current is low. With the shield bypass as shown in Figure 6.1 this current can be a problem at high frequencies.

If the direct-in signal is to be attenuated as a function of instrument gain using solid-state switches, the attenuator must be designed to avoid errors caused by switching resistance. If the on resistance is 100 ohms and varies 20 ohms with temperature, the resistors being switched should be greater than 20,000 ohms to avoid 0.1% errors.

The logic within a solid-state switch heats the pins that connect the switch to the circuit. This heating differs for the on and the off states for most switches. It produces switching errors at dc of about 10 μV.

Solid-state switches must be powered from sources driven by the input guard potential. Note this is not mandatory for relays. The switching logic must also function at the guard potential. This requires in turn that there be isolation between entering logic signals and the logic input to the solid-state switches. This isolation can easily be achieved through the use of optical isolators.

6.5 DIRECT INPUT MONITORING

Users often prefer to monitor input signals on a common bus. This allows a comparison of the amplified bus signal with the instrument output. The bus should be three wire to accommodate the normal differential input. This monitoring requires an input relay, and all of the comments made in Section 6.4 are applicable. Thermal gradients, relay coil power source, contact guarding, and so on, are all important factors.

Direct input monitoring has the following disadvantage. If the amplifier has failed and is upsetting the input signal, this fact may go undetected. In systems with adequate input calibration this problem can be avoided.

6.6 STRAIN GAGE CONDITIONING

The elements of strain gage conditioning are

1. Bridge excitation.
2. Bridge balance.
3. Bridge completion.
4. Calibration.

These functions can be housed with signal amplification in one instrument or housed as separate hardware items. Either approach is available in the instrumentation market. The advantage to placing the functions listed above into separate packages involves front panel area. The space available in a typical instrument is at a premium. The basic disadvantage is cost. Separate hardware requires separate cabinetry, separate connectors, and added interconnecting cables. Instruments are available that provide the necessary controls but with a few compromises in operability. Often bridge excitation is supplied separately so it can be used to excite several bridges in parallel.

6.6.1 Bridge Excitation

Excitation supplies should be grounded by the bridge input circuit. This circuitry is shown in Figure 3.3. If one signal lead is grounded at the source, then the isolation problems for the excitation source are significant. This is

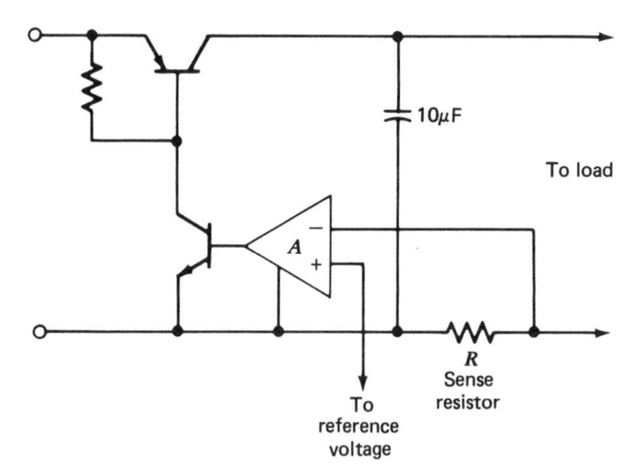

FIGURE 6.8 A constant current source.

discussed in Section 1.8.1. Grounding the excitation source rather than floating the source makes the best use of differential instrumentation and reduces cost.

Excitation sources can be either constant voltage or constant current. Output capabilities of 15 V and 100 mA are usually adequate. Constant voltage sources have output impedances below 50 milliohms and are usually bypassed by electrolytics so there is no sensitivity to reactive cable loading. Constant current supplies cannot be bypassed in this manner or the high-impedance output is violated. Most supplies are bypassed ahead of the sense resistor so that the reactance of this load can be increased by the feedback factor. If 10 μF is used and the feedback factor is 60 dB, this capacitor appears like 0.01 μF to the load. For bridge loads of 500 ohms this is a -3 dB frequency of 100 kHz. This technique is shown in Figure 6.8.

The loop gain of a constant current source should be kept high at low and midband frequencies or the resulting output noise will be excessive. Typical noise figures are 0.5 mV rms in 100 kHz bandwidth. The sense resistor should be large enough to produce a sense voltage maximum of about 1 V. For 100 mA this is 10 ohms. This robs the supply of about 1 V of compliance.

Most excitation sources can be switched from constant voltage to constant current with very little circuit change. The more difficult design problem is to guarantee absolute output stability for all loads, all voltages, and all currents.

6.6.2 Excitation Adjust

The manual adjustment of excitation level is straightforward. A multiturn potentiometer associated with the input reference voltage can set the current

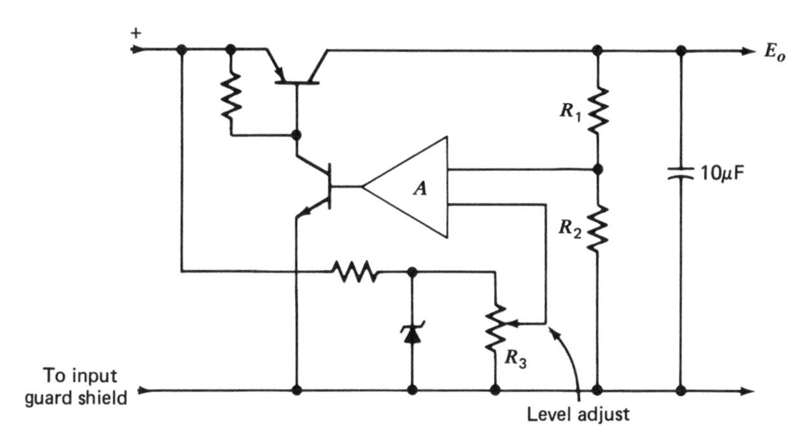

FIGURE 6.9 Excitation level adjust.

or voltage. The sensing circuitry is usually of high impedance to avoid problems when remote sensing is used. The attenuated reference voltage allows the voltage or current to be set to zero. This circuit is shown in Figure 6.9. If $R_1 = R_2$, the output voltage will be exactly double the voltage on the slider of R_3. Amplifier A should have an FET input.

The excitation level can be digitally controlled by supplying the slider voltages from a D/A converter. The digital input must again be isolated from the input guard-shield potential. This isolation can be expensive if the D/A binary input is carried between grounds in parallel on optical isolators. A 12-bit resolution would require 13 isolators—one isolator to strobe the data. A simpler method is to have two optical couplers. The first coupler carries a serial train of pulses to a counter. The second coupler carries a reset command to this counter. To change the counter memory, the counter is first reset and a known number of pulses is then transferred via the isolator to the counter. A parallel counter can be used to store the digital count at logic ground. This information can then be strobed onto a monitor bus as part of a computer-controlled instrument system.

6.6.3 Bridge Balance

A strain gage bridge is balanced when the bridge output signal is at null. To accomplish this, a balance potentiometer and a series SPAN R are connected as in Figure 6.10. When the potentiometer is at the top limit the SPAN R is shunted across the top bridge arm R_3. Similarly, the SPAN R is shunted across the bottom resistor R_4 when the slider is at the bottom limit. When the slider is at its center, the SPAN R provides no change to the balance.

The voltage for the balance potentiometer comes from the excitation source, but it could also come from a separate supply. When remote sensing

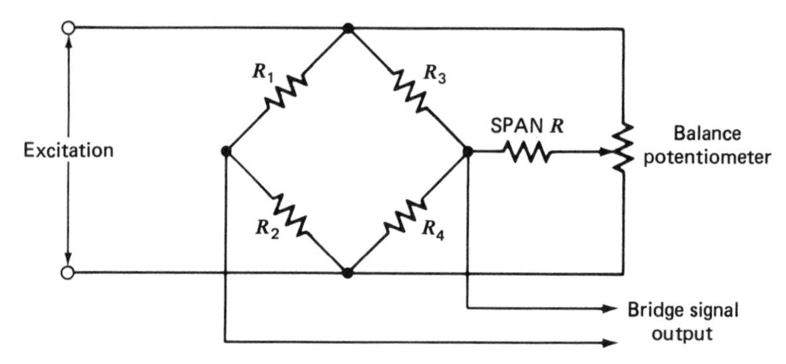

FIGURE 6.10 A simple balance control.

is used, this balancing voltage will be greater than the actual bridge excitation. This can be a slight problem on long lines where the line resistance may change with temperature. In this situation the balance may change slightly with temperature but cannot be blamed on the bridge resistances. With constant current excitation the balance potentiometer reduces the output impedance of the source unless the potentiometer is connected ahead of the current-sense resistor. When lines are long, an imperfect current source will allow line resistance changes to introduce an error.

The SPAN *R* resistor should be kept as high in value as practical. Since the SPAN *R* shunts the bridge, it does reduce the bridge sensitivity somewhat. When one or two active arms are used, these arms should be opposite the SPAN *R* side of the bridge.

6.6.4 Autobalance

The voltage supplied to the SPAN *R* can come from a digitally controlled source such as a D/A converter. This approach is used when an "autobalance" circuit is desired. Assuming that the SPAN *R* has been selected to be in the balance range, then the D/A converter counter must simulate turning the potentiometer. In finite steps the D/A converter increments the balance voltage until a comparison circuit indicates a zero crossing has occurred. At this time the clock supplying input to the D/A converter is turned off and the bridge is balanced.

The digital input to the D/A converter must be isolated from the input guard-shield potential. This requires that optical isolators be supplied for the D/A counter input—the reset command and the comparator signal. As in the earlier discussion, a parallel counter can be provided ahead of any isolation. This counter can be strobed to determine the balance voltage.

The balance resolution is determined by the full bit count in the D/A, the reference voltage, and the value of the SPAN R. If the D/A is 10 bits and the SPAN R accommodates a 1% bridge unbalance, the resolution is one part in 10^5. For a 10 V excitation level this is a resolution of 100 μV. If the SPAN R had a 0.1% balance range, the balance resolution would be 10 μV.

A higher resolution can be obtained using a 12-bit D/A or by prebalancing with a secondary manual system or both. If the prebalancing scheme is used, the autobalance SPAN R can be kept very high and a balance resolution down to a few microvolts is practical.

Amplifier noise limits the processes of bridge balancing. To balance a bridge the bridge output signal is monitored with a comparator for a zero crossing at the amplifier output. The noise that is referred to the input of an amplifier is typically 5 μV rms or 30 mV peak-to-peak at the output. This peak noise limits the resolution of the comparator.

One way to improve the balance resolution is to band limit the comparator so that the peak-to-peak noise is reduced. This in turn limits the clock rate that can be supplied to the D/A. If the comparator has a delay time of 1 ms, the D/A clock must be slower than 1 kHz to avoid overshooting the balance point.

For clock rates of 1 kHz and a count of 4000, it is obvious that several seconds are needed to achieve a balance. To avoid tying up external hardware during a balance cycle, it is desirable to have the autobalance process self-contained within the instrument. The only digital signals required to operate the autobalance circuit are a start command, a clock if 60 or 120 Hz is not used, and an alarm line indicating a "no-balance" condition.

More complex balancing systems can be built. For example, a fast clock can be used to get near a balance followed by a slow clock used to avoid balance overshoot. Other methods involve successive approximation or up/down counters based on comparator status. All methods should warn the user if the balance attempt was unsuccessful. Balancing should always reach some terminal status and not continue unchecked.

Dual-ranging autobalance systems can employ two D/A converters and two SPAN R's. In effect, the parallel manual control discussed earlier is replaced by a preliminary digital balance. Balancing first connects a D/A converter through a low value of SPAN R to the bridge. After the rough balance, a second D/A converter is connected through a high value of SPAN R to the bridge. If the ratio of SPAN R resistance is 1:10, then the first D/A might accommodate a $\pm$20% range and the second D/A a range of $\pm$2%. The voltages used in balancing can be either the excitation supply (proportional balancing) or a separate supply (ratiometric balancing).

Dual ranging is usually specified where solid-state strain gages might be used. Here large unbalances are apt to occur. The advantage of dual ranging is the reduced balancing time. If the rough balance has a 4-bit resolution and the final resolution is 8 bits, the maximum balance time is $2^4 + 2^8 = 272$ clock times. If a single 12-bit system is used, the maximum balance time is 4096 clock times.

Instruments frequently have filters to limit bandwidth. Comparator signals should be taken ahead of these filters or else the filters should be disabled during balance.

6.7 BRIDGE COMPLETION RESISTORS

It is preferable for bridge completion resistors to be soldered into the circuitry. Contact connections are apt to be noisy unless they are gold plated as in an edge connector.

Wire-wound resistors are almost a necessity for bridge completion elements. The balance point should hold to within 1 μV in 10 V, which is one part in 10^7. Very few resistor types can provide this level of stability and repeatability.

A SPAN *R* resistor may shunt a bridge arm by as much as 2%. This resistor needs only to be good to one part in 10^5. A metal-film resistor is perhaps acceptable, but a wire-wound resistor is often the user's choice.

6.8 CALIBRATION

Calibration resistors typically unbalance the active bridge to provide a near full-scale signal. If the full scale is 15 mV, a 10 mV signal is produced by a change in resistance of 0.4%. For a 150 ohm gage this is a shunt resistance of 37.5 kohms.

There are several philosophies regarding calibration. One approach implies a lack of confidence in the linearity of the instruments or recorders, and multipoint calibration is specified. With the quality of today's instruments a bipolar full-scale calibration is usually adequate. This method verifies the gain and also that the full-scale output is not restricted by an undefined load.

FET switches can be used to connect calibration resistors to the bridge. A FET switch with a 20 ohm "on" stability and a 10^8 ohm "off" resistance will not affect the calibration accuracy. The switches themselves must operate at the input or guard potential and should be optically isolated from logic ground.

If the calibration sequence involves a group of switches, it may be desirable to operate the calibration sequence by logic located in the input shield environment. In this case the only signals that need to be isolated are a start command and a clock to drive the sequencing logic.

If the calibration processes are handled by relays, the problems discussed in Section 6.4 are applicable. Relay coils may require guarding and all connections are potential thermocouples. If the calibration processes are short lived, then heating phenomena can be ignored. If someone is manually recording each calibration level, then the heating phenomena may pose a problem.

Multipoint calibration can include bipolar steps, an open excitation source, and a shorted input. These last two steps tell the user if the instrumentation is sensitive to source inpedance and also if the instrument is correctly zeroed. These last steps require relay contacts because FET switches do not have a low enough "on" resistance.

Instrument settling time must be considered in any calibration process. Any filtering can be automatically removed or the clock rate can be changed to accommodate the added settling time.

Calibration over long lines requires that separate leads be brought back from the active gage elements. This subject was covered in Section 3.1.5. The calibration lines can share the signal lines, but the leads supplying the excitation should not be shared. The instrument should be designed so that the user has the option of calibrating on either separate or common leads according to the application.

6.8.1 Series Calibration

In series calibration, the bridge elements are used as a resistance in an attenuator. The calibration resistor makes up the second element in the attenuator. The signal applied to the attenuator is the normal excitation voltage that is removed from the bridge. The instrument sees a signal that is proportional to the excitation level and to the bridge resistance. This method is useful where an active gage is not desired during the calibration process. The disadvantage is related to the switching necessary to reconnect the signal leads and the excitation. This method does not require separate calibration leads to avoid the IR drop in the excitation lines.

6.9 SIGNAL MONITORING

A common bus is frequently routed to all instruments in a system. The instruments can be monitored singly for excitation, input signal, or output level over this one bus. The switching of signals to this bus can be handled by remotely controlling relays or from a front panel switch.

The presence of a bus line connected to an undefined ground entering each amplifier can couple noise into each amplifier. The difficulty is compounded by the fact that the input leads are perhaps the most sensitive points in the instrument. The designer should be careful to shield this bus with the guard potential wherever it goes within the instrument. At the relay, the contacts can also be shielded by the techniques shown in Section 6.4.

The monitoring bus should not be grounded twice, that is, once by the signal and second by the monitoring instrument. This indicates that an instrument quality device should be used for monitoring purposes. The guard shield for the monitor should share the guard shield of the instrument being monitored. If an output signal is monitored, the output ground should be used.

6.10 MULTIPLEXING

Solid-state switches can be used to connect signal lines to a bus or to an instrument. When a group of signals is switched in a timed sequence, it is usually called a multiplexer. The switching of preamplified signals is usually called high-level multiplexing, whereas switching signals without preamplification is usually referred to as low-level multiplexing.

Solid-state switches have the obvious advantage of speed. The operate plus settling time is typically less than 1 μs. The on and off resistances can be accommodated by avoiding loads on the switches. Offsets are mainly due to thermocouple effects caused by logic dissipation within the switch.

A serious problem with low-level switching is the capacitive coupling of logic signals into the switching path. The logic signals have leading edge transitions that occur in nanoseconds. These leading edges couple through capacitances below 1 pF to the switch proper. A 5 V transition in 20 ns causes 25 μA to flow in 0.1 pF. This pulse current can flow toward both switch contacts. If it flows toward the source, it is attenuated by the source impedance. If the source is shunted by 1 μF, the pulse is reduced by the capacitance ratio or by a factor of 10^7. The current flowing toward the open contact is attenuated by a much smaller factor and can cause several volts to appear at an amplifier input.

The logic within the switch has its own delays. These delays allow leading edges to couple to the switch before the switch is actually operated. If the switch is being used to connect a signal to an instrument, a logic pulse will arrive at the instrument before the signal.

Various schemes are used to cancel the logic pulses discussed above. An equal and opposite pulse can be capacitively coupled to the switch to cancel the offending signal. This technique can reduce the pulse by a factor of ten. The instrument can be kept at a low gain until the pulse is attenuated. The disadvantage here is that both the instrument and the signal must settle to a final value before a sample can be taken.

If the pulses described above are ever rectified, the resulting signal can be invalid. Clamps are necessary for protection against unexpected overloads. If a clamp is used to limit pulse amplitude, rectification can cause an invalid charge to appear on the input capacitances and the sampled value will be incorrect.

The sampling amplifier must have excellent settling time characteristics after an overload. If gains are switched as a function of signal level, the gain settling time is also relevant. High-speed gain changing requires solid-state switches and these switches also couple logic pulses into the amplifier circuitry. Before a signal can be sampled, the gain and signal must settle to an acceptable error limit. This settling time determines the sampling rate of the multiplexer. Typical error limits are 0.02%.

Low-level multiplexers require that all three input signal leads be switched. This means that the guard shield potential must change for each signal. Signal and common-mode levels can vary over a wide dynamic range, which poses a problem with crosstalk. To reduce this phenomenon a supercommutation scheme can be used. This is shown in Figure 6.11.

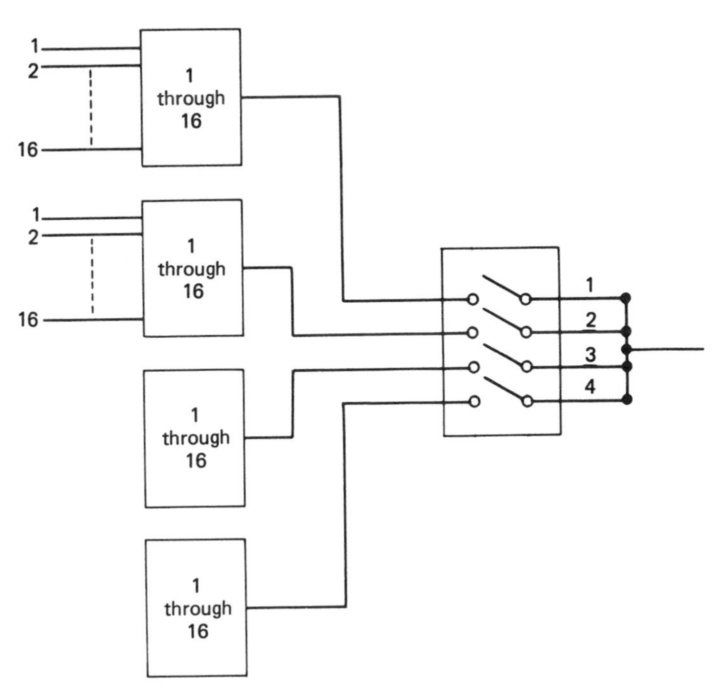

FIGURE 6.11 A supercommutation system for 64 channels. Each switch is a three-pole switch.

Each switch is a three-pole signal switch for a 64-channel multiplexer. The cross coupling can occur only from three other selected channels in the last group of four switches.

High-level multiplexers are not as sensitive as the low-level devices described previously. The channels are all grounded at one point and therefore a single switch is adequate for each signal. An active device being sampled should not be sensitive to the sampling pulses or errors can result. Sensitivity might include ringing because of an inductively reactive output impedance.

6.11 ALIASING ERRORS

Sampling processes can easily go astray if signal filters are not supplied. Aliasing errors are best described by an example. If a 1000 Hz signal is sampled at 1 kHz, the result will be a dc signal whose amplitude depends on the phase relationship between the signal and the sample time. If the sampling frequency is 1001 Hz, the sample point will "walk" along the 1 kHz wave. In one second the set of sample points will describe a 1 Hz sine wave. This is obviously not the signal being sampled. To avoid this class of error a filter must be used to remove signal content above one-half the sample rate.

If the signal is known to be free of frequency content above one-half the sample rate, then filters are not required. This may be difficult to guarantee if the results are not already known. Filters are often an expensive item in a system; one common practice used to avoid aliasing errors is to increase the sample rate. This allows the use of a simpler filter or an available filter.

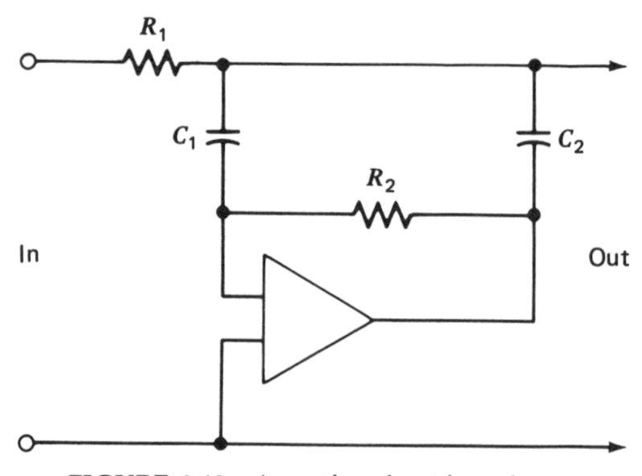

FIGURE 6.12 An active shunt impedance.

6.12 ANTIALIASING FILTERS

Digital filters cannot be used to remove aliasing errors. The example given above where the sampled output is dc illustrates this point. No amount of sophistication will eliminate this dc value once it is sampled.

To optimize usable bandwidth, antialiasing filters can be rather sophisticated. The steep cutoff character found in Chebyshev or elliptical filters can be used. Tables are available for informing the designer the type of filter needed to meet a given specification.

In high-level multiplexing, filters are generally a part of each signal amplifier. This is acceptable assuming that the prefiltered signal has not overloaded the electronics.

Low-level multiplexing poses another set of problems. Filtering usually consists of series resistors and shunt capacitors. This circuitry not only filters the signal but it also provides a low-impedance source for the multiplexer switches. Obviously, this filtering does not provide a very steep cutoff character. The resistor values are limited to about 1000 ohms to avoid errors caused by amplifier source currents.

For small signals polar electrolytic capacitors can be used as filter elements. Voltages should be held below 0.5 V to avoid nonlinear loading. These electrolytics are not used for accuracy but for small size.

Active low-level filters can be built using a concept known as an active impedance. The circuit shown in Figure 6.12 fits this description and provides a second-order transfer function for the signal E_{in}. Note that the filter adds no dc component to the signal. The impedance looking back into the filter output is very low and is ideal for feeding a multiplexer.

Circuits similar to that shown in Figure 6.12 can be built in balanced form and cascaded to form higher-order filters.

CHAPTER SEVEN

AMPLIFIER CONSIDERATIONS

"Cloudy mornings turn to clear evenings."

The methods available to the designer for rejecting common-mode signals can be grouped as (1) derived-guard, (2) guard-driven, (3) transformer decoupling, and (4) optical decoupling. These methods vary, depending on available components, operating needs, specifications, and on the economics of design.

Within the instrument the designer must position filters, offsets, signal conditioning, power supplies, output stages, gain switching, and so on. There are many choices in circuitry and in circuit layout. Within these many choices the designer hopes to optimize performance and reduce cost.

The zero stability provided by a design is of primary concern. In multiplexing systems a zero signal input channel can be used to rezero the instrument during each cycle. In other instruments a "chopper" is used for continuous drift correction. Most high-gain designs rely on the balance stability of an input pair of transistors. Each method has its merits and proper application.

Gain blocks needed to provide functions such as filtering, offset, variable gain, high output current, input balance, and common-mode rejection are provided by the ubiquitous IC (integrated circuit). These gain blocks are relatively inexpensive, consume low power, and use up little space. The problem in design thus reduces to configuring these gain blocks and providing the end product with a satisfactory set of controls.

7.1 THE DIFFERENTIAL GAIN BLOCK (DGB)

An integrated circuit can be configured to amplify differential signals and reject common-mode signals (see Section 1.7). This basic circuit is shown in Figure 7.1. If the resistors are in the ratio $R_1/R_2 = R_3/R_4$ then

$$E_o = (E_2 - E_1)\frac{R_2}{R_1} \tag{7.1}$$

If $E_1 = E_2$ then $E_0 = 0$ and the circuit rejects the common-mode signal.

To provide gain in this circuit the resistor R_2 can be connected to an output divider as shown in Figure 7.2. If $R_3/R_4 = R_1/R_2$ and if

$$\frac{R_5R_6}{R_5 + R_6} + R'_2 = R_2 \tag{7.2}$$

then E_0 is given by

$$E_o = (E_1 - E_2)\frac{R_2}{R_1}\left(\frac{R_5 + R_6}{R_6}\right) \tag{7.3}$$

The circuit in Figure 7.2 can be used to both attenuate and then reject a common-mode signal. If $R_3/R_4 = 10$, then the common-mode level $(E_1 + E_2)/2$ can be 100 V. Note that the signal level at the amplifier is only 10 V. If the ratio $(R_5 + R_6)/R_6 = 10$, the net overall gain is unity. Obviously, gains greater than unity are attainable.

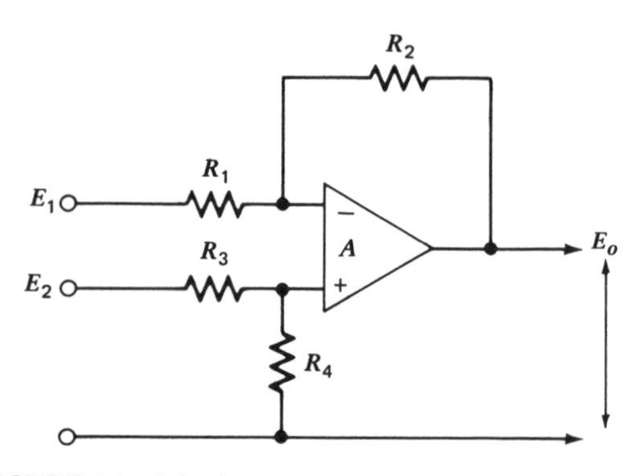

FIGURE 7.1 A basic common-mode rejection amplifier.

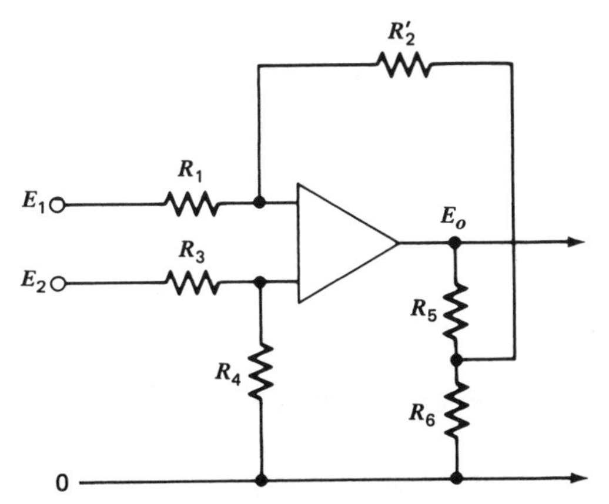

FIGURE 7.2 Gain in a common-mode rejection amplifier.

7.1.1 Applying the Differential Gain Block

The input impedance to the circuit in Figure 7.2 is approximately R_3. Depending on bandwidth requirements, this value may be typically 20 k to 100 kohms. This is not a high enough impedance to satisfy the needs of a low-level instrument (see Section 1.6) This circuit is adequate for isolating 1–10 V signals between two ground points. This application is shown in Figure 7.3. To maintain a wide bandwidth and high input impedance, resistance R_1 can be made up of a series of three or four resistors. This keeps the voltage gradient low and the effective shunt capacitance low. If common-mode rejection ratios greater than 60 dB are desired, the resistors should be wire-wound and resistors R_2 should be made adjustable over a small range.

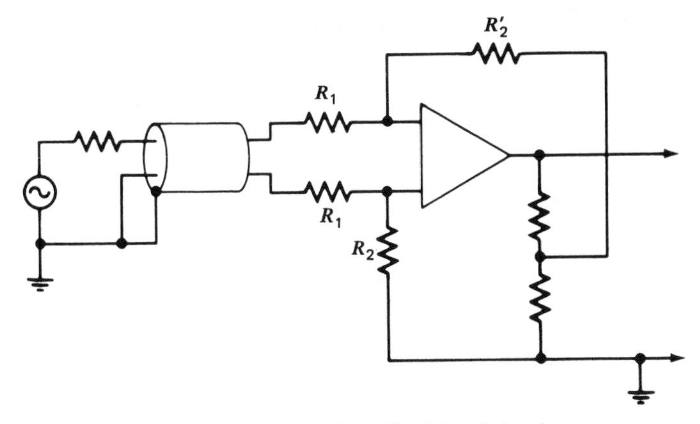

FIGURE 7.3 Isolation of high-level signals.

The circuitry shown in Figure 7.3 is located at or near the point of signal termination. If several high-level signals terminate on one ground, a group of these amplifiers with a common power supply can be effectively used to reject ground potential differences.

7.1.2 High-Impedance Isolation

The input impedance of the circuit in Figure 7.2 can be raised by adding isolation buffers. The buffers must operate at the source ground potential and this requires a separate power supply. This circuit arrangement is shown in Figure 7.4.

Here two differential gain blocks are used. The first block using the resistor ratio R_1/R_2 removes the source common-mode potential between E_1, E_2, and E_{g1}. (See Section 1.7 and Figure 1.11.) The second differential gain block removes the ground-to-ground common-mode signal $E_{g1} - E_{g2}$. The unity-gain FET amplifiers A_1 and A_2 provide a high input impedance for the signal leads. Resistances R_3 and R_4 allow a path for ground current to flow but not for current in the signal path. Resistors R_3 and R_4 attenuate the ground difference common-mode signal and determine the peak value that can be rejected. This type of circuit is usable for gains less than 100. Any additional gain can follow the circuitry shown in Figure 7.4.

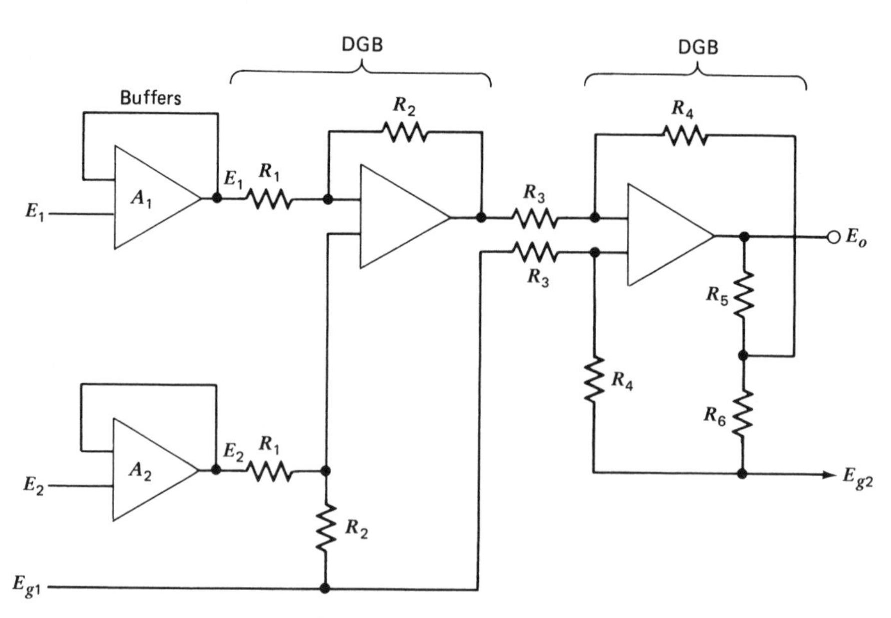

FIGURE 7.4 Isolation buffers.

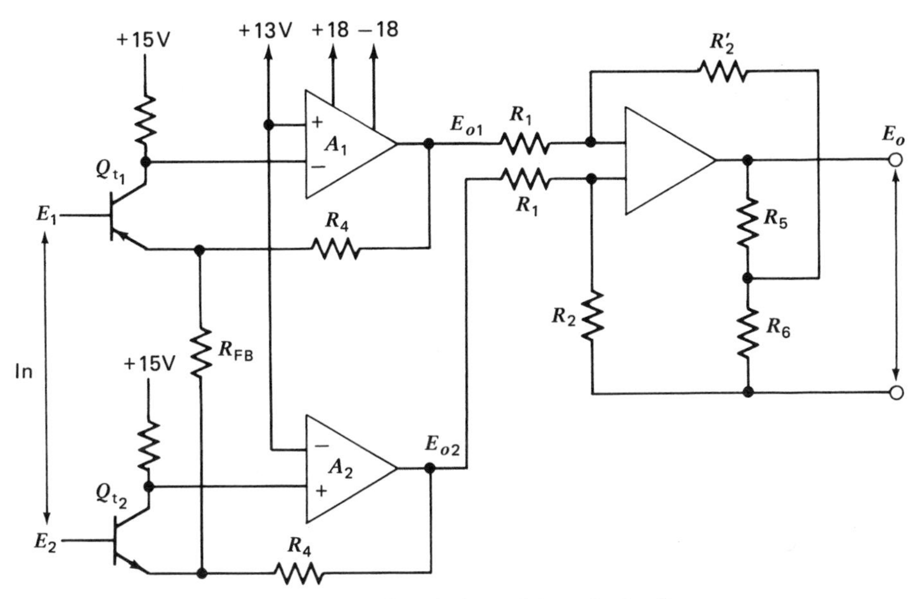

FIGURE 7.5 A typical transistor pair circuit.

7.2 HIGH-GAIN DIFFERENTIAL AMPLIFIERS

The input circuitry shown in Figure 7.4 is limited in performance by the matching qualities of A_1 and A_2. Even though A_1 and A_2 may be located on the same dice, the drift and noise qualities are not the same as a matched pair of transistors. This topic was discussed briefly in Sections 6.1 through 6.3.

To achieve a high input impedance and use a matched transistor pair, the circuit in Figure 7.5 is usually used.* Tranistor Q_1, amplifier A_1, and feedback resistor R_A form a feedback loop identical to the loop formed by transistor Q_2 and amplifier A_2. The only link between these two loops is resistor R_{FB} where FB is feedback. If $E_2 = 0$, the emitter of Q_2 is held fixed. The loop consisting of Q_1 and A_1 has a feedback path with an attenuator formed by R_A and R_{FB}. The gain to point E_{o1} is thus

$$E_{o1} = E_1 \frac{R_A + R_{FB}}{R_{FB}} \tag{7.4}$$

*This configuration is not unique. See *Linear Data* book LF152 published by National Semiconductor Corporation, where current feedback is used.

Similarly

$$E_{o2} = E_2 \frac{R_A + R_{FB}}{R_{FB}} \tag{7.5}$$

The gain to the signal $E_{o1} - E_{o2}$ is simply equation 7.4 minus equation 7.5, or

$$E_{o1} - E_{o2} = (E_1 - E_2)\left(1 + \frac{R_A}{R_{FB}}\right) \tag{7.6}$$

Note that when $R_{FB} = \infty$, the gain is simply unity. If $E_1 = E_2$, $E_{o1} = E_{o2}$, which is independent of the value of R_A or R_{FB}. Because the emitters follow the input bases, the input impedance is maintained high.

When $E_1 = E_2$, the signal $E_{o1} = E_1$. This results because R_{FB} returns to a voltage $E_2 = E_1$. Similarly, $E_{o2} = E_2$. In other words, $E_{o1} - E_{o2}$ represents the difference input signal with gain equal to $1 + R_A/R_{FB}$ and $E_{o1} + E_{o2}$ represents the common-mode signal $E_1 + E_2$ with a gain of unity. One advantage to this circuit is that the differential gain can be varied by changing one resistor. A disadvantage is that the lowest gain possible is unity.

The difference signal $E_{o1} - E_{o2}$ in Figure 7.5 is followed by a differential gain block. In this way the difference signal is referenced to an output ground as a voltage E_o. Any common-mode signal $(E_{o1} + E_{o2})/2$ is rejected by the gain block (see Section 7.1). The magnitude of the common-mode signal is dependent on the resistor ratio R_2/R_1.

7.2.1 Power Considerations

The power supplied to amplifiers A_1 and A_2 can be derived in two ways. The first method is a floating power source driven by a derived common-mode signal. (See Figure 6.2 and Section 6.1.) The second method is a separate power source driven by the input shield. The derived common-mode signal allows the user to misuse the input shield and still provide for common-mode rejection.

The floating supply can be derived from a separate transformer or from a pair of stacked constant current sources. This circuit is shown in Figure 7.6. The voltages E_3 through E_6 are used to operate the input circuitry including A_1 and A_2. These voltages all move with the common-mode signal derived by the summing resistors R_{CM}. The ± 15 V supply is used to power amplifier A_3 and any subsequent stages of gain. In this circuit the derived common-mode signal can swing approximately ± 50 V. This limitation occurs when the signal voltage plus the common-mode voltage plus the supply volt-

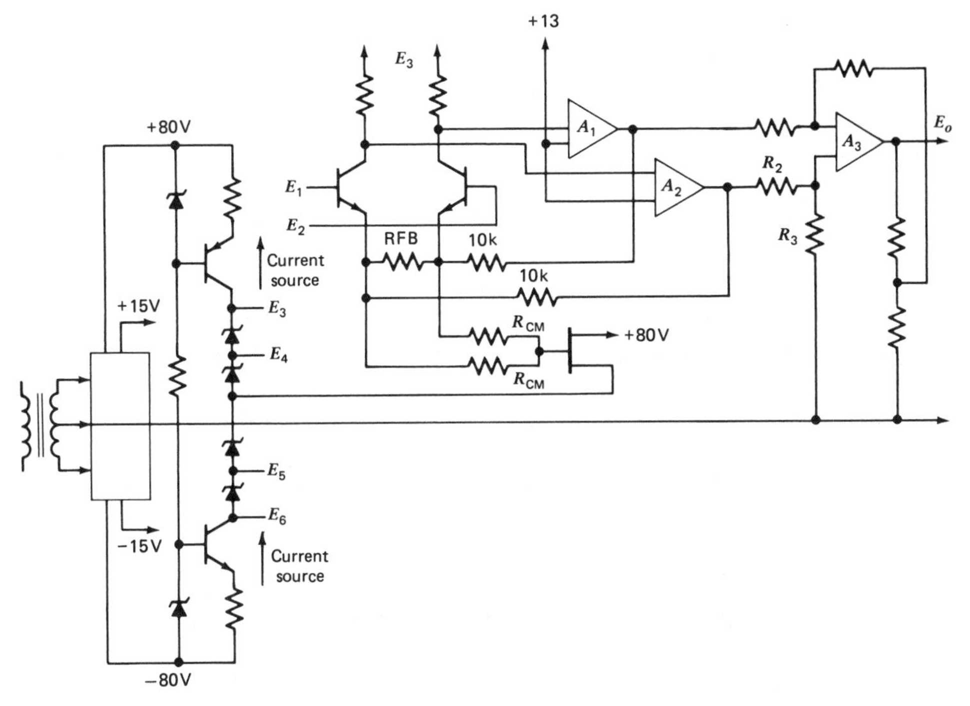

FIGURE 7.6 A stacked constant current source.

age E_3 reduce the collector voltage on the constant current source to near zero. Of course, the ratio R_2/R_3 must also be adequate.

If the common-mode voltage is 300 V, then the constant current sources would need to be powered from ±330 V supplies. These high voltages add shock hazard to the design, increase the cost, and add to the internal heat dissipation.

The separate transformer approach is shown in Figure 7.7. Transformer T_1 supplies power to the input transistors and amplifiers A_1 and A_2. The secondary has an average potential determined by the input guard connection. A separate transformer is used to operate the output stages. Instead of two transformers, one transformer with two center-tapped coils can also serve to power the two circuits.

The circuit in Figure 7.7 can accommodate high common-mode voltages without the drawbacks of the circuitry in Figure 7.6. The resistor ratio R_1/R_2 must be high enough to attentuate the common-mode signal to acceptable limits. The secondary coil of transformer T_1 must have a breakdown voltage rating sufficient to accommodate operating voltages plus common-mode signals.

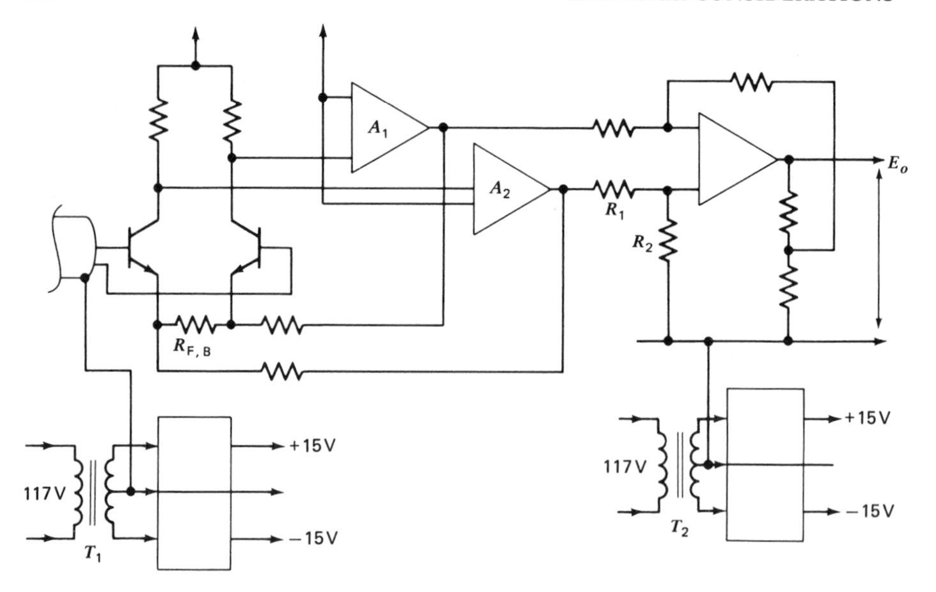

FIGURE 7.7 A separate transformer supply approach.

7.3 HIGH-VOLTAGE DECOUPLING

The circuits shown in the last section all had a resistive path between the input and output stages through the basic differential block. This path allowed the common-mode potential to force some ground current flow at dc. To avoid this current flow and to increase the permitted common-mode levels, transformer or optical isolation can be used.

A transformer can be used to carry modulated signals across an electostatic barrier. If the modulator and the demodulator operate at 100 kHz, the resulting bandwidth can approach 30 kHz. The linearity of such a modulator/demodulator is limited by the performance of switches and the characteristics of the transformer. Some designers have operated the modulator in a constant current mode to reduce coil voltages and improve linearity. The advantages of modulator/demodulator scheme as stated before are high common-mode voltage levels and no low-frequency current paths between circuits. A block diagram of this scheme is shown in Figure 7.8.

The designer must be careful that modulation currents do not circulate back through the source and reenter the amplifier via the input. This can result if the transformers are improperly balanced or shielded. Constant current operation helps to remove this difficulty. Any recycled signal modulation causes signal distortion and gain error. Simple carrier feedback causes offset errors.

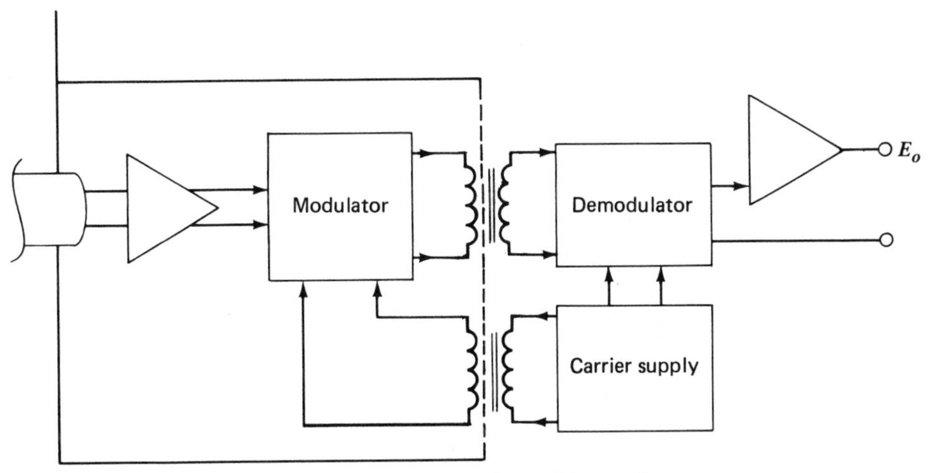

FIGURE 7.8 A modulator/demodulator scheme.

This circuitry has the disadvantage of being more complex and requiring separate power supplies on each side of the isolating transformers. If the demodulated and filtered signal is not buffered, then an output power supply is not required. In this case, the carrier signal can be derived in the input guard environment.

7.3.1 Optical Isolation

A signal that is converted to digital form by a suitable A/D converter can be transmitted over an optical link. This technique truly breaks the ground connection and can accommodate very high common-mode voltages. Other than the cost and accuracy considerations the approach is straightforward. Because an optical path can carry gigahertz data rates, many channels can be multiplexed over one such link. If each channel is sampled at 10 kHz with 10-bit accuracy, then 100 channels require a bit rate of only 10 MHz.

Optical isolators can be used as analog feedback elements to achieve high voltage isolation. Consider the circuit shown in Figure 7.9. The voltage E_o' will adjust until the current in isolator diode D_1 causes current I_1 to flow in Q_1 equal to E_{in}/R_1. This forces the summing point voltage E_s to zero. If a matched optical isolator D_2 causes the same current to flow in Q_2 then $I_2 = I_1$. If $R_4 = R_2$, then E_o is a direct measure of E_{in}. Stated another way, the nonlinearity in the forward optical isolator is canceled by placing a matching isolator in the feedback loop. This circuit must be duplicated if both signal directions are to be accommodated. Performance is limited by the availability of matched isolators over a useful temperature range. Isolators of this type

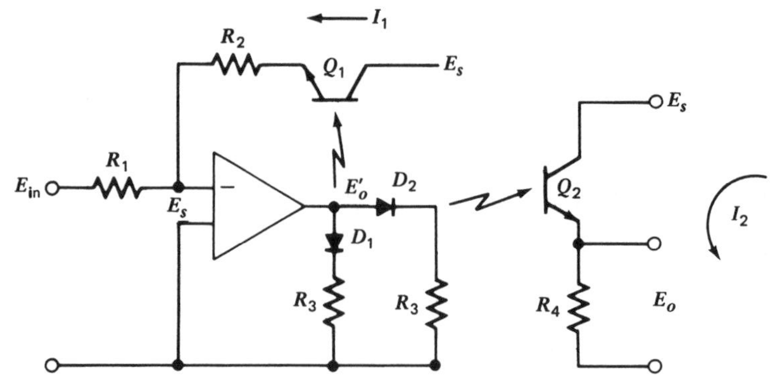

FIGURE 7.9 Optical isolation and feedback.

can easily withstand a 1000 V common-mode voltage for short periods of time.

7.4 COMMON-MODE SPECIFICATIONS

The discussion on common-mode processes in Section 1.6 suggested a rejection ratio of 120 dB at 60 Hz for a 1 kohm line unbalance. This in turn requires leakage capacitances of less than 2 pF. This performance usually degrades 20 dB per decade as a function of frequency and gain. At 600 Hz and at a gain of 100, the rejection ratio might be reduced to 80 dB. A lower gain is usually associated with higher signal levels. A proportionately lower common-mode rejection ratio (CMRR) therefore maintains a constant signal-to-error ratio.

CMRRs are always measured as an input error signal. For example, if a 10 V common-mode signal carries a 10 mV error at the output of a gain 100 instrument, then the referred-to-input error is 100 μV. The rejection ratio is 10 V/100 μV or 100 dB.

Common-mode rejection processes can take place anywhere along the gain path. Gains can be switched before or after this point. The CMRR varies with gain, depending on where the gain is switched and where the rejection error takes place. If the rejection error is generated at the instrument input, the CMRR is independent of gain. If the error is generated postgain, the CMRR will fall off with gain setting. An as example, in the circuit of Figure 7.5 the gain precedes the differential gain block and the CMRR falls off with gain setting.

Common-mode specifications for instruments are almost always limited to the rejection of ground potential differences. Common-mode signal with respect to input guard is rarely specified. Quality performance requires both specifications to be about the same.

Another specification that is rarely discussed is common-mode slew rate. The gain to common-mode signals in Figure 7.5 between E_1 and E_{o1} is unity. This means this circuit can respond to 10 V common-mode signals while amplifying millivolt signals. If slew-rate limits are exceeded, signal rectification results and an unwanted conversion to differential signal takes place. Note that a distortion of 0.1% can produce a full-scale output.

7.5 GAIN SWITCHING

Gain changes in the circuit shown in Figure 7.5 can be accommodated by changing one resistor, R_{FB}. If $R_A = 10$ k, then R_{FB} equals 10 ohms for a gain of 1000 (see equation 7.6). For a gain sequence of 1, 2, 5 . . . there are two approaches to gain switching: (1) a separate value of R_{FB} for each gain step or (2) a value for each decade of gain and a separate 1, 2, 5 sequence located elsewhere in the design. This latter approach uses fewer resistors for seven or more gain steps.

Gain changing should occur as near to the input as practical. This optimizes drift and noise referred to the input. The rule is simple: The noise or drift of a stage in an instrument is reduced by the gain preceding that stage. If a stage with a gain of one precedes additional gain, then the noise of the first and second stages will add together in an rms sense. If the first stage has a gain of three, then the combined noise will only be increased by 5%. Drift may not add in an rms sense and therefore the second stage drift will be reduced only by a factor of three.

The 1, 2, 5, sequence can be made a part of the differential gain block or an added stage. If the differential gain block is used, precision resistors are necessary to maintain a high CMRR. If a post 1, 2, 5 gain stage is used, the gain following the differential gain block will reduce the CMRR by the stage gain.

It is possible to build the 1, 2, 5, sequence into the differential gain block so that a single CMRR balance control is practical. This approach requires a separate buffer amplifer A as shown in Figure 7.10. This buffer provides a low impedance to connect the 1, 2, 5 attenuator to feedback resistor R_2.

The circuit in Figure 7.10 can be changed to include a variable gain control. This control might provide a gain multiplier of one to three for gain steps of 1, 2, 5. A disabling switch can be included to allow the user to obtain the

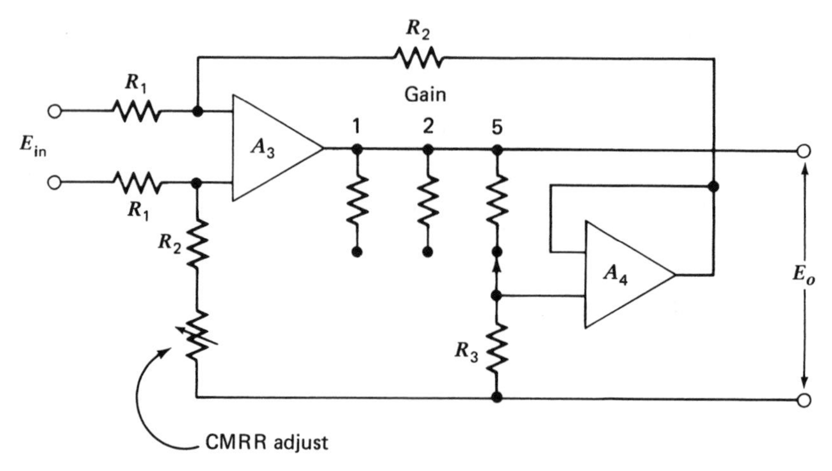

FIGURE 7.10 A buffer amplifier used in the feedback loop.

exact gain multiplier steps. Such a variable gain circuit is shown in Figure 7.11. The disable switch connects the amplifier to the top of the variable-gain potentiometer. The gain trim adjust allows for inaccuracies caused by the potentiometer loading of resistor R_3. The CMRR adjust allows the resistor ratios R_1/R_2 to be accurately matched. To optimize both CMRR and gain stability the resistors should all be wire wound.

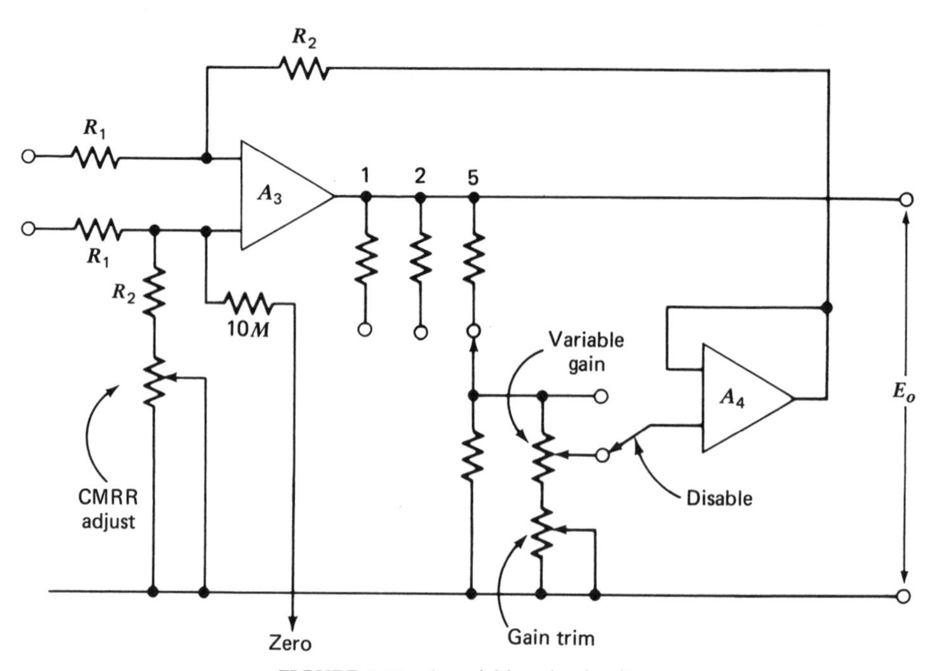

FIGURE 7.11 A variable gain circuit.

7.5.1 Gain Attenuation

Gains below unity are often required in a general-purpose instrument. If this attenuation follows the input stage, the maximum output voltage will be limited to the unity gain level times the attenuation ratio. If the amplifier is normally 10 V full scale, an internal 10:1 attenuator will limit the peak output level to 1.0.V. In this same example a balanced preattenuator allows a 10 V full scale output signal and a 100 V input signal.

To be practical the input attenuator cannot match the 1000 megohm input impedance of the instrument. A comfortable resistance value for 10:1 attenuator might be 1 megohm and 111 kohms. Even here a 1 pF parasitic capacitance across the 1 megohm resistors will start to peak the response above 100 kHz. Small capacitors across the 111 k resistors can compensate for this effect.

7.5.2 Gain Accuracy

Instruments are frequently specified as having 0.1% gain accuracy. This accuracy is usually measured at low frequencies under no-load conditions with a low source impedance. This accuracy requires 0.025% gain-determining resistors and some form of gain trim for the entire instrument. Square-wave balancing techniques are preferred or sine-wave testing for measuring gain (see Section 10.1).

Some users rely on a calibration source and the variable gain control to set gain. This technique allows the output voltage to be read directly in engineering units. As an example, if a transducer is calibrated at 10 lb = 13.26 mV, then the user sets the variable gain control so that a 13.26 mV input causes a 10 V output. The user now interprets each output volt as a pound. This forces the user to reset his gain if he changes transducers. This application places no demand on gain step accuracy and a 1% value is adequate by default.

7.6 ZEROING ERRORS

Instruments have zeroing errors that are referred to the input (RTI) or referred to the output (RTO). An RTI error is gain sensitive and must be multiplied by the gain of a device before its effect can be compared with an RTO error. RTO errors are not gain sensitive and usually dominate at low gains. In adjusting for zero the RTI errors should be removed first. The adjustment is correct when the output zero is unaffected by gain changes. RTO adjustments are made last.

Ideally RTI correcting signals should be introduced at the instrument input. To avoid input loading, RTI corrections are usually made after the input stage. The zeroing might involve varying the collector resistor of Q_1 in Figure 7.5.

The differential gain block shown in Figure 7.5 follows the input gain stages. When this block is switched in gain as in Figure 7.11, then this circuit has its own RTI and RTO errors. If subsequent circuitry is not switched, then this RTO error can be included with all other RTO errors and removed by one adjustment. The RTI error can be removed by summing a small current into the positive input of amplifier A_3. Again, this error is balanced out when changes in gain produce no change in output zero.

7.6.1 Temperature Drift

The matching of transistors Q_1 and Q_2 in Figure 7.5 determine the RTI drift specifications with temperature. Pairs are available with offset drift of less than 0.1 μV/°C. In applications requiring additional temperature compensation, a separate temperature-sensitive element is needed. This element can be as simple as a diode or a high-temperature coefficent resistor. This element allows a controlled temperature-dependent current or voltage to be added to the signal path. To avoid an offset the compensation signal can be balanced by a nontemperature-sensitive signal. Figure 7.12 shows a simple balancing circuit. When control R_1 is at its extremes, a positive or negative temperature-dependent current flows in R_2. Control R_3 adds a nulling current through resistor R_4. As the temperature changes, this circuit can provide a variable temperature-dependent signal of either polarity.

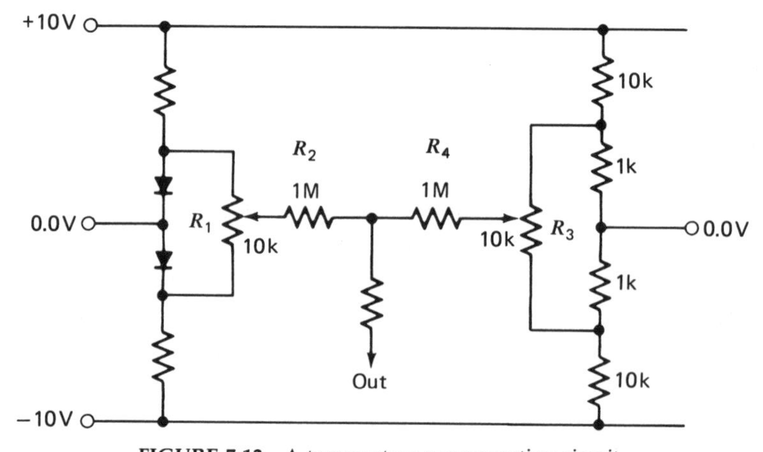

FIGURE 7.12 A temperature compensation circuit.

It is necessary that the temperature-sensing elements be mounted on or near the elements causing the drift. The point of compensation and the remaining circuitry are not critical as to location. The sensor location is critical if over or under compensation is to be avoided.

7.7 OFFSET

It is sometimes useful to offset an amplifier to accommodate signal range. For example, if the signal is unipolar, the amplifier can be offset to one extreme and the output voltage swing can be doubled. This approach is viable if one signal extreme is at zero volts output.

Assume a ±10 V maximum output swing. For an input signal varying from 90 to 100 mV, the maximum gain without overload would be 100. The output of the instrument would vary from 9 to 10 V. If the RTO offset were made −10 V, the output signal at gain 100 would vary from −1 to 0 V. At gain 200 the output would cover the ranges of 8 to 10 V. In this case the RTO adjustment increased the dynamic range by a factor of two.

To further improve the dynamic signal range the amplifier would have to be offset RTI. If the RTI signal were 105 mV, the effective input signal range would be ±50 mV. At a gain of 1000 the output covers the range ±5 V. This is a tenfold improvement over the case with no RTI offset.

An RTO offset can be included in a design with little difficulty. This is not the case for RTI offset. Any circuitry that involves the input must not add drift or reduce CMRR. The RTI signal itself must be drift free.

The least expensive method of adding RTI offset is with a small series battery. A simple attenuator can be used to set the offset voltage. If a 1.35 V cell is rated at 80 mA hours and an offset of 100 mV is desired, the life expectancy is 500 hours. Batteries require maintenance and the contacts are subject to corrosion. For most applications this approach is unsatisfactory.

An RTI offset signal can be generated by associating two constant current sources with a series input resistor. This circuit is shown in Figure 7.13. The ±15 V supplies must be referenced to the derived guard potential or to the input guard shield. The two current sources must be equal and controlled from a common signal. Note this circuit provides a unipolar RTI offset signal. If connections to the resistor are reversed, the RTI signal is also reversed. The entire circuit in block diagram form is shown in Figure 7.14. This circuitry illustrates the difficulty of supplying RTI offsets to an instrument. For this reason this feature is seldom offered in this form. These constant current sources can be set by a D/A converter and this allows for remote control of RTI offset.

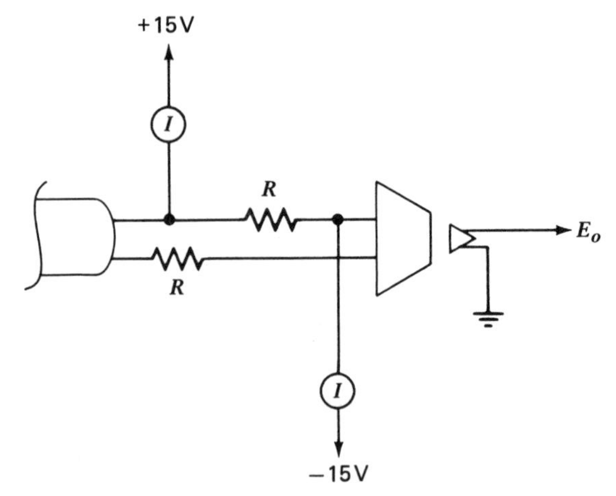

FIGURE 7.13 An RTI offset circuit.

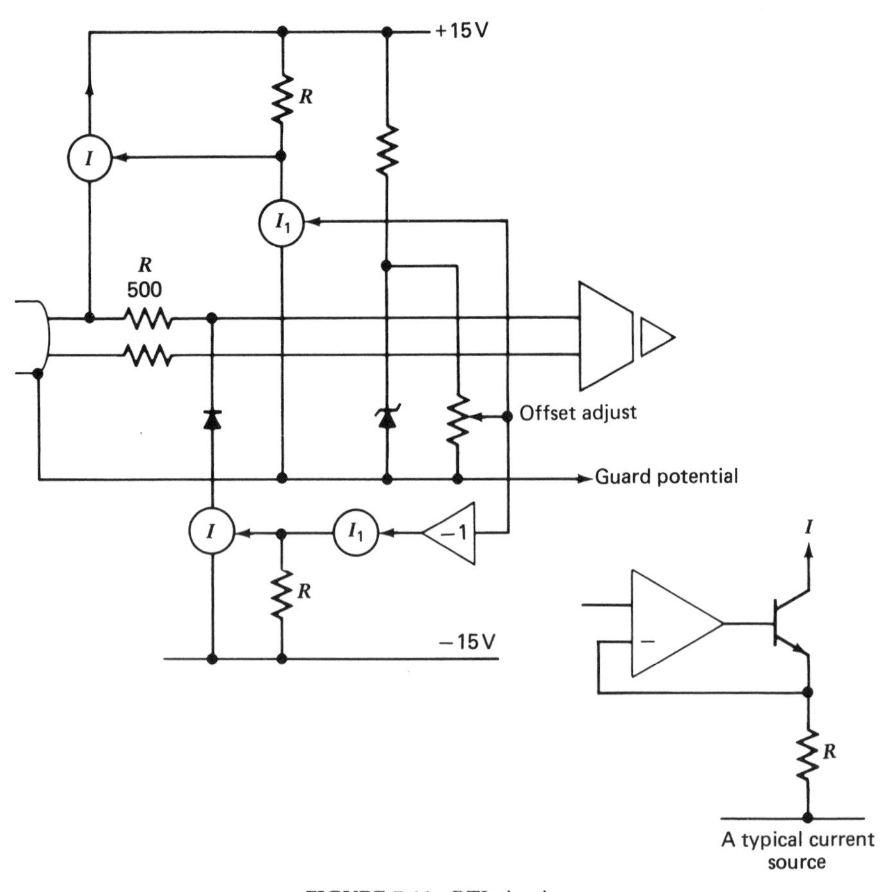

FIGURE 7.14 RTI circuitry.

RTI offsets can also be generated by varying an added emitter resistor. See figure 7.5. This method usually adds drift as a function of temperature. The battery source described in the preceding paragraph can be replaced by a floating power supply derived from a few turns on a transformer. If shielding is handled properly, this source can be as effective as a battery.

RTI offsets that function by forcing current through the source impedance and return through the guard conductor are less complex. Here the RTI offset voltage is dependent on the source resistance value. For long input lines the voltage drop in the input conductors must be considered. This offset signal is usually unipolar and the input leads need to be reversed to reverse the RTI offset sense.

7.8 MULTIPLE OUTPUTS

Designs can easily provide several outputs. These outputs can be independently filtered, provided with variable gain, offset, or be voltage limited. These parameters can be switch selected or hardwired into the instrument. The output impedance of these outputs is usually below 0.5 ohm. A paralled RL circuit is usually placed at each output to stabilize the circuits against reactive loading.

Multiple outputs must share a single signal return unless separate power supplies are provided. For short signal runs this poses no problem, but note

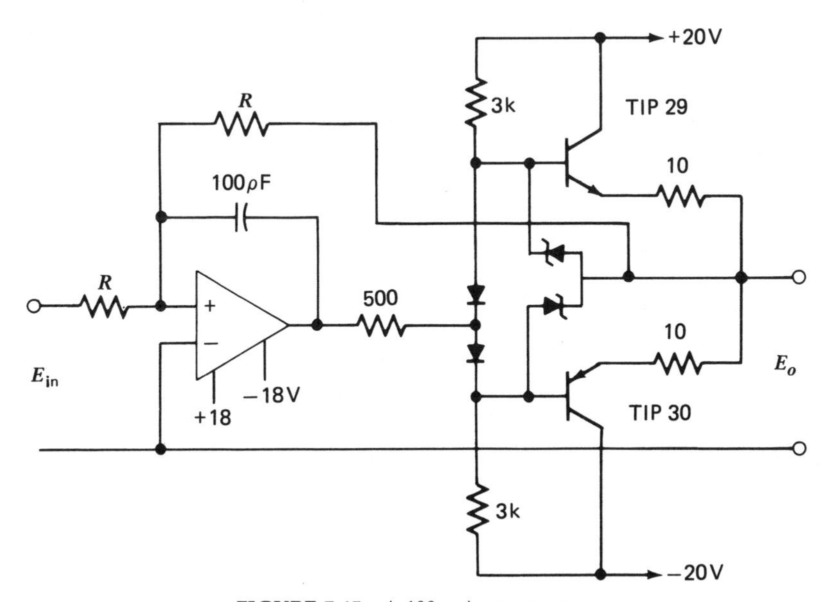

FIGURE 7.15 A 100 mA output stage.

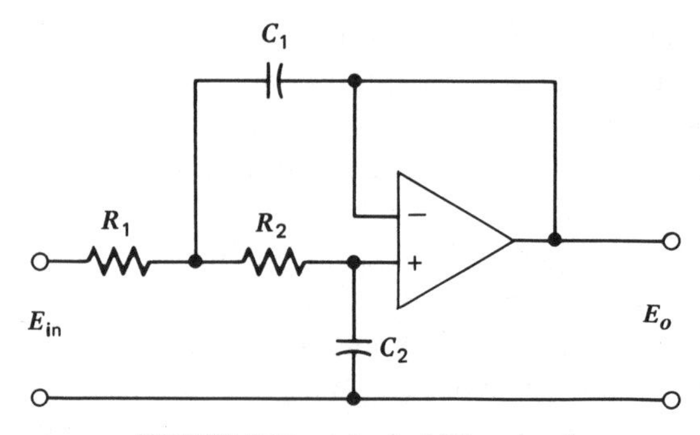

FIGURE 7.16 A typical filter circuit.

that terminating outputs on separate grounds will create output ground loops. Load currents can take multiple paths to return to the instrument and this can cause cross coupling particularly at high frequencies.

High-current output stages (±100 mA) are available as single integrated circuits. They can also be built out of an IC and discrete components. This circuit is shown in Figure 7.15. The zener diodes act as current limiters. The capacitance around the amplifier keeps the circuit stable by avoiding the delays in the power transistors.

7.9 FILTERING

Filtering can be provided in a positive unity-gain amplifier. A typical circuit is shown in Figure 7.16. The transfer function is equal to

$$\frac{E_o}{E_{in}} = \frac{\omega_0^2}{s^2 + 2\delta s\omega_0 + \omega_0^2} \tag{7.7}$$

The natural frequency is equal to

$$\omega_0 = \frac{1}{2\pi\sqrt{2R_1R_2C_1C_2}} \tag{7.8}$$

The damping factor is equal to

$$\delta = \frac{1 + a^2}{2ab} \tag{7.9}$$

where $a^2 = R_1/R_2$ and $b^2 = C_1/C_2$. If $R_1 = R_2$, then

$$\delta = \frac{1}{b} \tag{7.10}$$

A third-order filter can be formed by preceding the circuit in Figure 7.6 by a simple R_3C_3 filter. A convenient value for $R_3 = 0.1R_1$. The component values can be found by first deriving a circuit transfer function in terms of C_1, C_2, and C_3 assuming $R_1 = R_2 = 1$ and $R_3 = 0.1$. The coefficients of s in the desired transfer function are compared with the corresponding coefficients in the circuits transfer function. A trial value for C_2 leads to values for C_1 and C_3. In the product relation $C_2C_2C_3 = K_1$, a new value for C_2 is calculated. The initial value for C_2 is varied until the derived value agrees with the assumed value. After a suitable set of capacitor values are selected, these values are scaled to accommodate reasonable impedance and frequency requirements.

Switch-selectable filters are a common feature in instruments. It is practical to switch resistor values over a two decade range. By using an FET amplifier, the frequency range of 10 Hz to 10 kHz can be covered using only two sets of capacitors and switched resistors.

Signals buried in noise can overload an instrument prior to filtering. A filtered overloaded signal can lead to disaster. This would indicate that prefiltering is a necessity. One solution is to place the last factor of ten in gain after the filter. This improves the dynamic range by 20 dB, but obviously reduces the CMRR and the RTO drift.

Another solution is to split the filter. In a three-pole filter the first pole can be placed directly at the input. The remaining two poles can then be located at the output. The input pole may reduce the unwanted signal to the point where postfilter gain is not required. If this solution is still insufficient, the two-pole section can be incorporated into a gain-ten amplifier.

Higher-order filters can be placed directly at the input of an instrument without adding drift. This circuitry is discussed in Section 6.12.

CHAPTER EIGHT

INSTRUMENT REQUIREMENTS

"Tall trees catch much wind."

8.1 INSTRUMENT HOUSINGS

Instruments are available on the market as individual plug-in modules, as multisignal modules, and as unpowered gain blocks. The individual modules are usually housed in multichannel rack modules (cabinets) provided with a set of rear panel connectors. These rack modules can be fan cooled or convection cooled by vertical air flow. The multisignal packages are often a part of a larger system where front panel controls per channel are minumum or where the major functions are computer controlled. The unpowered gain blocks are often treated as ICs and are functionally imbedded into other circuits using power supplied to the card. These ICs are often listed as instrumentation amplifiers by their manufacturers.

The individual amplifier modules are usually housed for protection in a metal or plastic case that has removable side covers. Any metal outer wrap is insulated from the rack module and is frequently connected to the input guard shield. Mechanical card protection is usually ignored by digital card designers and there has been a trend in low-level instrumentation to copy this approach. The lack of shielding is acceptable if the designer takes proper care of input leads and a few sensitive components.

A 300 V common-mode specification allows the input guard shield to be 300 V above output ground (rack cabinet ground). Theoretically this voltage can be found on front panel screws, knobs, and so on, unless these are all intentionally insulated or protected. In most applications the 300 V signal appears infrequently and little danger exists.

The desire to keep the cost and size down on larger systems has brought about 1 in. thick instrument modules (16 across per 19 in. rack module). These modules can have their own transformers or can be powered from common supplies. If common supplies are used, many performance specifications are often compromised. For example, if the output leads are short, then output ground loops cause no problem.

The edge connectors used extensively in digital circuitry are now used regularly in analog instrumentation. This approach removes the cost of an added connector in the instrument module. The fingers on the pc board and the contacts on the edge connector should be gold plated to accommodate low-level input signals and high-impedance connections. Tinned connections are apt to become unreliable over extended periods of time.

Adjacent fingers on one side of the board are recommended for input connections. This reduces loop area and places both leads in the same thermal environment which reduces thermocouple effects. These fingers should be surrounded by connections to input guard shield to reduce unwanted mutual capacitances.

It is practical to bring 117 V through these edge connections. A small fuse resistor can be used at the fingers in series with the power line to protect the board in case of an accidental short. This resistor can be 10-ohms-⅛-watt carbon. It is preferable to hardwire transformer primaries and not use board traces connected to the utility power.

8.2 COOLING

Individual rack modules (cabinets) can be built for free vertical air motion or individually fan cooled. The use of fans creates a set of problems. The fan pulls air out through the rear surface so that if unfiltered air enters each instrument, dirt can accumulate in each module. This forces the need for a regular cleaning schedule that is often difficult to maintain.

The vertical free flow approach works as long as the temperature rise in the uppermost equipment is within bounds. If the air at the entry point is refrigerated, then moderate vertical heating is not a problem. This air motion should be continuous and at one temperature. If the temperature cycles, then every input connection will change temperature but at different rates. The resulting thermocouples will cause apparent input drift in the instruments.

8.3 CONNECTORS

The cost of cabinetry can be greatly reduced if intermediate cabinet wiring can be eliminated. An interfacing connector mounted in the cabinet that receives the instrument on one side and cables on the other side would be ideal. However, this arrangement is not yet available at the time of this writing.

Input connectors are usually three pins for signal or ten pins if a full bridge is to be connected. Ideally the shell of each input connector should float and be connected to the input guard shield. A plastic rear panel provides this float without the use of insulating washers.

Although output connections are shown in this text as two-wire shielded cables, little performance is lost if a two-wire connection is made. This two-wire connection can be coaxial or even a simple twisted pair. BNC connectors are in common use for output connections.

The one case where an edge connector is not satisfactory is the input to a charge converter. A separate connector is required to maintain a tight shield. The highest gain has an equivalent charge noise of 0.001 pC. For a 1 V contaminating signal, the leakage capacitance must be below 0.001 pF, a value impossible to hold except through a solid connector.

8.4 CONTROLS

Front panel controls to change amplifier parameters are not readily abandoned. A great deal of sophistication is required to let a computer set the functions and give the user a "hands-on" feeling of control. Most users cannot or will not pay the price.

Controls define most of the cost in instrumentation. If there are no controls, there is no need for a front panel or its screening. Instead of a gain switch, gains can be set by soldering in place single resistors. Internal controls can set RTI and RTO offsets. Filter components can likewise be soldered in place. What is left is the basic block of gain, which represents only 15% of the cost in a typical instrument.

Small switches and potentiometers have made it practical to place a large number of controls onto a very small panel area. The limitation is frequently the panel area that is required to title each control according to function and position. Controls are frequently recessed screwdriver adjustments, for small knobs are clumsy and often changed by knob-twisting passersby.

Internal DIP switches (miniture pc switches) can be used to set operating modes. When multiple outputs are supplied, the switches can select variable gain, offset, or filtering for one or all outputs. Again, the switches can be omitted and any particular configuration can be hardwired into the instrument.

8.5 ACCURACY

Certain specifications have become standards over the years. A common-mode rejection ratio (CMRR) of 120 dB at gain 1000 with 1000 ohms unbalance is just one such milestone. Gain accuracies of 0.1%, linearities of 0.01%, noise figures of 2 or 3 μV-rms RTI in 100 kHz bandwidth are all fairly common. These specifications form one basis for discussing system accuracy. This book, however, has set out to show how the entire signal path must be considered before any performance figures can be discussed. A CMRR specification loses its meaning with an improper shield connection. A 100 kHz bandwidth is not guaranteed over a long cable just because the instrument has this bandwidth. In other words, a specification is a measure of instrument performance and not a measure of system performance. The attitude of many users is to base system performance on instrumentation specifications and ignore many basic system errors.

Any group of errors can be handled on a statistical basis. The error bounds and a confidence figure can be calculated. This approach is valid for well-defined parameters measured under known conditions. This error analysis must consider each contaminating influence including its magnitude and bandwidth. An error study can be complex and time consuming and even if completed may not hold for a next test.

8.6 PHASE SHIFT

Instrumentation amplifiers with bandwidths that include dc have phase shifts that increase with frequency. If the instrument is free from transient overshoot, this phase shift will increase linearly with frequency to at least 50% of the 3 dB point. Any phase shift represents a time delay at that frequency. If the phase shift is linear, then all sinusoids are delayed an equal amount of time.

Instruments are usually made up of several cascaded sections, with each section contributing to the overall phase shift. Gains are usually controlled in one section by changing feedback resistors. A gain change results in a change of bandwidth and a subsequent change in phase character. These changes can be reduced if the bandwidth is also controlled in the feedback structure.

For many measurements the time delay through an instrument is important. To be correct, the time delay should include the transducer and the input cable. If active filters are included in each instrument, this filter can serve as the dominant time delay. Obviously, all filters must be set to the same cutoff frequency if the delay is to be the same through each channel.

Time delay is a function of filter cutoff frequency, filter type, and filter order. Control of cutoff frequency is not sufficient to define time delay. The best measure of time delay is phase slope or $d\phi/df$ where ϕ is the phase angle and f is frequency. This slope is most affected by the least-damped poles in the filter. An adjustment on this one damping factor can serve to match time delays between channels.

The standard low-pass filters such as Butterworth, Bessel, and Chebyshev are mathematically convenient. These names serve to communicate to the user the full nature of the filter. These filters optimize one parameter, either the frequency response or phase response, in a prescribed manner. None of the standard types optimizes both phase linearity and amplitude flatness in the pass band. This type of optimization can be handled by a computer. The author has coined the term PILAF filter (Phase Inband Linear Amplitude Flat) to describe a filter designed in this manner. None of these filters is unique and thus there is no simple way to characterize them as a family. Fortunately, any design can be completely described by its transfer function.

It is surprising how well a PILAF filter can perform. A fourth-order design can hold the amplitude flat to within a few percent at one-half the 3 dB point, maintain a linear phase shift within the pass band, and have less than 1% step-function overshoot.

8.7 ALARMS

The user of an instrumentation system prefers to be warned if anything goes wrong before or during a test. A variety of these alarm conditions can be sensed and brought out to a LED for viewing or to an alarm bus for centralized reporting.

Alarms can be provided for strain gage operations. An open bridge can be sensed for a no-current condition. An autobalance circuit that does not balance can also provide an alarm. In some applications alarms can be based on threshold signal values, whereas in others an overscale signal might trigger an alarm.

Alarm conditions can be sensed in the input guard environment or in the output circuitry. These signals should not be connected directly to external logic or ground loops will result. The safest solution is to use optical couplers. An alarm bus must enter each instrument and thus can act as a source of crosstalk or signal contamination. Care must be taken in routing and shielding these alarm lines.

CHAPTER NINE

DIGITAL CONTROL

"Of two evils choose the least."

The digital or computer revolution will continue its impact on instrumentation. Logic ICs, optical isolators, and FET switches are digital components that are well established in analog instrumentation. The real impact is yet to come as labor costs soar, the number of channels increase, and data analysis gets more sophisticated. The future is pointing toward more and more computer control of instrument functions such as setting operating parameters and initiating calibration and balance. Other functions include output level monitoring, configuration control, and alarm status.

9.1 THE DIGITAL INTERFACE

Digital ground potentials can cause ground loops when connected to instrumentation grounds. This forces the use of isolation between the various instrument grounds and digital common. Optical isolation is preferred because of its small size and low cost. Relays can be used in less sensitive areas.

One approach to digital control is to place a microprocessor in each instrument. A command and data bus would operate the instrument and let a remote device interrogate or monitor the instrument. This microprocessor would obviously sit idle most of the time.

The binary signals required to set gain or excitation levels must be coupled into the input guard region. An optical line per binary bit is not economical and therefore it is desirable to use a serial path of some sort. The data can be set into a counter one significant bit at a time or simply counted in as a long train of pulses after a reset. It would appear that a microprocessor is again an overkill.

The problem seems to be best solved by controlling many instruments through one controller, which in turn connects to a computer. An instrument recognizes a position-oriented address and responds to a set of commands. The commands simply direct clock pulses to various counters (memories) that control the instrument's functions. The counters that follow the optical isolation can be paralleled by a counter on the controller side. With this arrangement all settings can be interrogated, as shown in Figure 9.1. Note that only eight lines are needed to interrogate or change the settings of 16 different functions. Each function requiring optical isolation requires a minimum of two optical couplers. However, high-level functions that can be controlled by relay contacts may not require optical coupling.

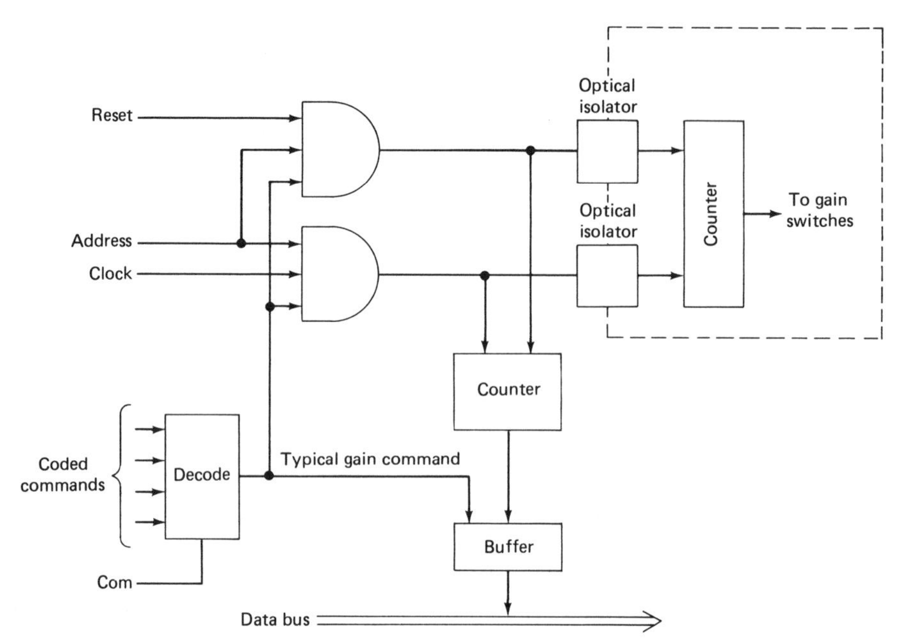

FIGURE 9.1 A typical command logic scheme.

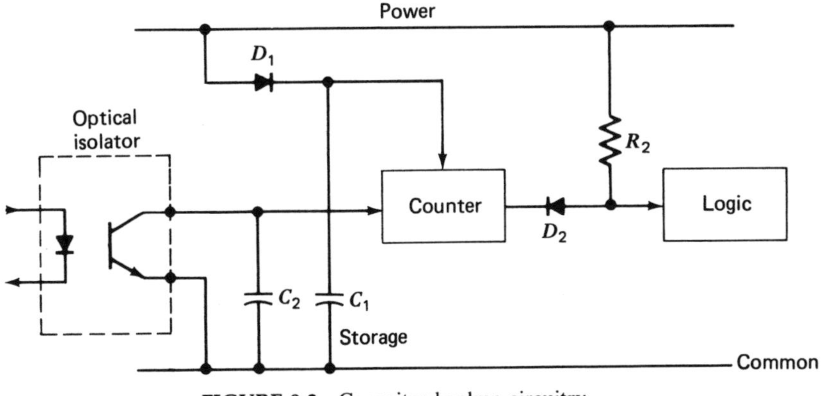

FIGURE 9.2 Capacitor backup circuitry.

9.2 BATTERY BACKUP

The logic described in the previous section can be designed using low-powered CMOS (Complimentary Metal Oxide Semiconductor). This logic class requires nanoamperes of standby current. Diode steering allows the memory elements (counters) to remain active even when the main power is removed. The standing power can come from storage capacitors or small rechargeable batteries. At 1 nA a 10 μF capacitor will lose half of its voltage in about 12 hours. For most servicing problems or for retaining memory during a power failure, the capacitor storage is adequate.

To reduce unwanted power drain, the logic must not sink or drive current to other ICs with loads. This isolation can be provided by using decoupling diodes with pull-up resistors in appropriate places. It is also necessary to avoid transient reset during any "power-up" process. This can be done by delaying the reset signal. Typical diode steering circuitry is shown in Figure 9.2. Diode D_1 keeps the charge on C_I from draining back into the supply. Diode D_2 removes any load imposed by logic connected to the counter. Capacitor C_2 slows the reset signal so that start-up transients will not reset the counter. This delay need be only a few microseconds if it is required.

9.3 CALIBRATION

Calibration processes can involve direct signal substitution or pseudosignal generation. Direct signal substitution may use a relay to reconnect the instrument input to a known signal on a calibration bus. Pseudosignal generation can unbalance a bridge by using either shunt calibration resistors or adding RTI offset.

The RTI offset can be controlled digitally by using a D/A converter associated with a binary counter. The signal resolution is defined by the number of bits on the converter and the counter.

Calibration by signal substitutions simply requires an appropriate signal relay. This relay can be latching or nonlatching and can be controlled by decoding command functions. Because the relay is associated with input circuitry, optical isolation is recommended.

Bridge calibration schemes are numerous. The variables include one or more resistors shunting one or more adjacent bridge arms, pairs of resistors shunting opposite bridge arms for unipolar or bipolar signal generation, open excitation, shorted input, or combinations of the above. With the use of today's very linear instruments, multipoint calibration is often not required.

The calibration resistors associated with wire gages are usually above 30 kohms in value. This means they can be connected using FET switches. The "open" resistance of a FET switch is above 10^7 ohms and the "on" resistance is below 100 ohms. When the gages are solid state, the gage elements are apt to have a high resistance and FET switches may not be adequate.

The particular calibration scheme that is selected can be initiated by one command and the sequence can be automatic. If the sequencing clock is externally controlled, a fully automatic system can be stopped at each point in the sequence. This method does take sequencing logic within the instrument. Commands can also be allocated to operate individual relays or FET switches so that calibration sequences are controlled by external logic. This short discussion indicates the large number of calibration schemes that are possible.

9.4 COMPUTER ANALYSIS AND VERIFICATION

A computer-controlled system can be changed during a test. This includes changing gain ranges, filter settings, offset, and so on, in a fixed time sequence or as a function of signal history. This latter type of operation requires real-time data processing. All changes must be carefully logged by the system as a function of time so that final data will be meaningful.

Data analysis requires multiplexers and associated antialiasing filters. The controlling algorithms must be carefully considered so that system changes result in benefits not in inadvertent losses of data.

Computer control allows for a full status report on each instrument just prior to a test. This includes all functional settings and some form of calibration. These same tests can be made just following a test to further validate the fact that all signal paths remained intact during the run. This type of verification could take days on a manual basis but seconds through using a computer. A validation test after a run need only report discrepancies or changes that occurred during that run.

CHAPTER TEN

TESTING AND CALIBRATION OF INSTRUMENTS

"Lost time is never found again."

A rapid and accurate testing routine for instruments is a necessity when hundreds of channels are involved. Unless the routine is carefully devised, many testing errors can result. Testing for 0.1% performance using precision signal sources and meters can be very tedious and often requires much interpretation. The methods described here are simple and accurate provided a few rules are followed.

10.1 NULL TESTING

When a signal is first attenuated by a factor $1/G$ and negatively amplified by a factor G, the sum of the resulting signal and the original signal should be unity. Any departure from unity represents an amplifier error and is a measure of amplifier performance. This null circuit is shown in Figure 10.1. For this circuit to function correctly, resistor R_2 must be wired as a four-terminal element. If $R_2 = 10$ ohms, then each milliohm of resistance in series with R_2 represents an 0.01% error.

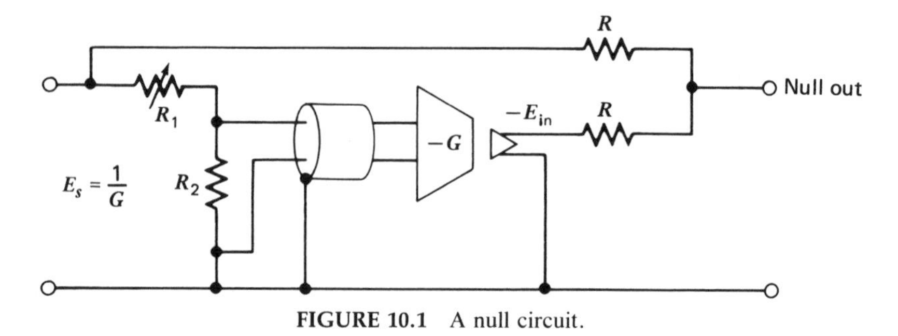

FIGURE 10.1 A null circuit.

The signal E_s can be any waveform, but sine waves and square waves yield the most information. If sine waves are used, the null error will be dominated by the phase shift in the amplifier unless very low frequencies are used. The summing resistors R need to be matched to within 0.01% and should preferably be precision wire-wound elements. The null point measures any amplifier error divided by a factor of two. A detected null error is multiplied by two and compared with E_s to measure the error as a percentage of signal.

10.1.2 Square-Wave Testing

To avoid the issue of phase shift which, after all, is not an error, null testing is far more effective if square waves are used. If the attenuation and gain in Figure 10.1 are exactly equal, the only time a null signal will not occur is when the amplifier is in transition. If the amplifier responds in 1 μs, the null error for a ±10 V square wave will be ±5 V spikes lasting 1 μs. These spikes will occur at every square-wave transition. This testing is thus independent of square-wave frequency and has no corresponding phase shift errors.

The null point can be viewed on an oscilloscope where signals below 1 mV are easily viewable. The ±5 V spikes can sometimes overload the oscilloscope input and thus clipping diodes are usually placed at the null point to limit the spikes to ±0.6 V. The null error signal for a typical square-wave test is shown in Figure 10.2.

10.2 POSITIVE AND NEGATIVE GAIN

The null test shown in Figure 10.1 works for a negative gain amplifier. If the amplifier is a differential instrument, the input leads can always be configured for negative gain. For positive gain configurations the summing re-

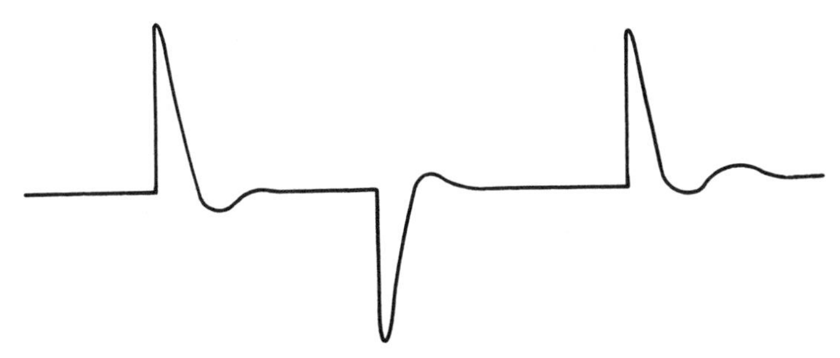

FIGURE 10.2 A typical square-wave null signal.

sistors are omitted and the monitor is placed between the signal source and the instrument output. This test circuit is shown in Figure 10.3.

The oscilloscope cannot be third-wire grounded. The null signal is a direct measure of the error. Note that oscilloscope overload protection requires a series resistor to avoid shorting the signal sources during transistion.

10.3 GAIN MEASUREMENT

The attenuator $1/G$ in Figure 10.1 can be made of two precision resistors. If the sensing resistor R_2 is 10 ohms, the value of R_1 at null is a measure of the gain G or

$$G = \frac{R_1 + 10}{10} \tag{10.1}$$

For example, if R_1 is 9,981 k, the gain is simply $(9{,}981 + 10)/10 = 999.1$. If a decade resistor box is used for R_1, the gains can be read directly from the dials. For gains below 100, R_2 should be raised to 100 ohms to avoid loading the signal source.

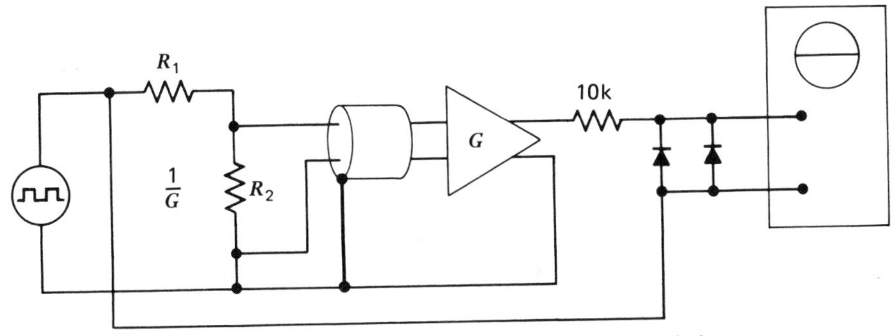

FIGURE 10.3 A square-wave null circuit for a positive gain instrument.

The summing resistors in Figure 10.1 can also be adjusted to obtain a null. In the example above, if the nominal gain is 1000, then R_1 is set to 10 (G-10), or 990. The summing resistor would then need to be lowered by 0.09% to achieve a null. With a properly calibrated potentiometer, gain variation can be read directly in percent. Two controls can serve to read ±1% and ±0.1% full-scale gain errors.

It might be argued that a square-wave test of gain measures the gain at only two output levels. Although this is true, if the linearity is better than the gain accuracy there is no problem. To further guarantee that the gain measurement is valid, the square-wave amplitude can be varied over all values and the null point should not change.

Gains can also be measured using low-frequency sine waves. The frequency should be low enough to avoid any significant phase shift. Any resultant null signal is a measure of linearity error.

Null-type measurements sidestep problems associated with zero errors. If the instrument has an offset, the null point is unchanged even though a static value is present in the observation. If meters are used to measure input and output signals to obtain gains, all offset errors must be subtracted from the readings.

10.4 SETTLING TIME AND SLEWING RATE

The null pattern shown in Figure 10.2 demonstrates the settling time of an instrument. Assume the full-scale input signal is 20 V. When the transient error signal reaches a ±10 mV band, the instrument has settled to within 0.1% of final value. It is necessary that the settling time of the summing resistors be better than the instrument.

For a linear instrument, the settling time should not vary as a function of signal level. When the settling time does vary, the instrument is said to be slew-rate limited. This simply means that the output voltage reaches a limiting transition rate measured in volts per second. Note that slew rates need not be equal in both directions.

Frequency versus amplitude response measurements that place the instrument into a slew-rate region are invalid. The slew rate of a sinusoid in volts per second is given by

$$E' = 2\pi f E_p \qquad (10.2)$$

where E_p is the peak signal and f is the frequency. If a slew-rate limit is reached at 10 V peak and 20 kHz, then 10 V signals cannot be used to measure the amplitude response above 20 kHz. In this example a 200 kHz response can be measured using a 1 V peak signal.

10.5 LINEARITY

A null pattern using low-frequency sinusoids gives a direct measure of linearity. The null pattern is adjusted for a minimum peak-to-peak value. A Lissajous pattern with the drive signal on the horizontal axis is often easier to interpret. Here all errors are measured with respect to the zero input signal point. This linearity measure is equivalent to a best straight line fit through zero output.

10.6 OUTPUT IMPEDANCE

Small changes in gain resulting from a level change are a measure of output impedance. Square-wave signals are recommended. If a 100 ohm load causes a 0.1% gain change, the output impedance is 0.1 ohm resistive.

Null techniques are not easily applied to measuring output inductance. Using sine waves, the simplest test involves finding the lowest frequency where 3 dB of additional attenuation results when a known load is placed on the amplifier. Care must be taken to avoid slew-rate limiting or current overloads. The inductive reactance is equal to the load resistance at this frequency and this defines the inductance.

10.7 INPUT IMPEDANCE

If a shielded resistance is added in series with an input lead, the resulting gain change measures the input impedance. For example, if the added resistance is 1000 ohms and the gain changes by 0.1%, the input impedance is 1000 megohms. This test is valid only on the negative input lead. To measure the positive input lead the positive gain null circuit is required.

The input capacitance of an instrument is best measured by using sinusoids and noting changes to the -3 dB frequency when resistors are added to an input lead. The resistance value must include any internal resistors that might be present to filter unwanted high-frequency signals.

10.8 PHASE SHIFT

The phase lag in the instrument can be canceled by adding phase lead to a subsequent circuit. When the summing resistor connected to the amplifier output is shunted by a small capacitance, this leading phase angle can be introduced. The phase shift of the instrument is then simply

$$\phi = \tan^{-1}\frac{X_c}{R} \qquad (10.3)$$

where X_c is the reactance of the capacitor and R is the value of the summing resistor. This technique is limited to a few degrees unless more sophisticated corrections are made.

10.9 AC GAIN

Square-wave null tests through an ac-coupled instrument will result in a waveform as shown in Figure 10.4. The sloped lines between the pulses result from the dominant ac-coupling element. The null is correct when the two slopes intersect in the middle of the transient. Sinusoidal null measurements are restricted to a 0° phase shift frequency. The square-wave null is not dependent on square-wave frequency, assuming that the instrument being measured has a flat midband frequency-versus-amplitude response.

10.10 COMMON MODE

Common-mode specifications often call for 60 Hz and thus testing involves this frequency. The input leads and input guard are all connected together and driven sinusoidally with respect to output common. Any resulting output signal is divided by the instrument gain and compared with the driving signal. This ratio is the CMRR. A resistor is then added to the two input leads one at a time and the CMRR is remeasured. The added resistance is usually 1000 ohms. This is the definitive test for it measures the leakage capacitance out of the input guard shield.

Common-mode rejection testing is not limited to 60 Hz, however. Specifications often require performance testing at much higher frequencies. Common-mode signals can cause internal slew-rate limiting. Testing of rejection ratios should not be mixed with testing of common-mode slew rate.

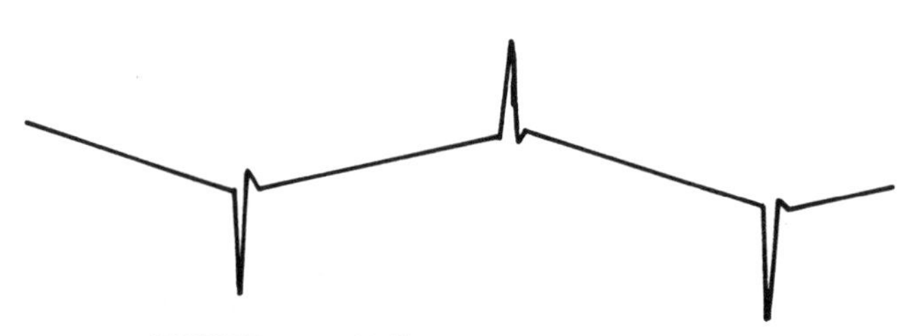

FIGURE 10.4 Null pattern in an ac amplifier.

If the CMRR is level dependent at high frequencies, then a slew-rate condition probably exists.

Square-wave testing of CMRR provides a better broadband measure of the rejection processes. Low-frequency problems will appear as sag in the error signal. Error-setting times measure the high-frequency performance. If the settling time is level dependent, then again there may be slew-rate difficulties.

10.11 VOLTAGE AND CURRENT MAXIMUMS

The maximum voltage or current capability in an instrument can be measured by noting the signal level where the gain null breaks up. This measure can be made with sine or square waves. It is easier to distinguish the polarity of the limitation using sine waves.

Square-wave testing can show up other problems. If the settling time varies widely with load, this indicates that an instrument cannot deliver its maximum current at high frequencies. If the recovery processes are constant with load changes, this indicates there is apt to be no added distortion due to loading.

10.12 LINE VOLTAGE TESTING

Null testing provides a convenient way of observing the exact point at which an instrument fails to function properly. Gain is measured at maximum signal voltage and maximum current load. The line voltage is then reduced until the null pattern breaks up. The breakup of the null defines the lower ac voltage limit of operation.

High line voltage can cause overheating, component breakdown, circuit instability, transformer core saturation, or added output noise. Line voltages should not exceed the manufacturer's recommendation. The above performance difficulties can be difficult to spot. For example, the instability problem may only appear on certain gains and then only with reactive loads.

Line voltage testing can involve 50–400 Hz sources, inverters, portable generators, line transients, and rf entry. Extensive testing in these areas is seldom done unless a specific problem is expected. Nonstandard power sources pose problems in several ways. They often cannot supply peak current demanded by capacitor input filters. Waveforms are often poor and the added harmonic content can affect instrument performance.

CHAPTER ELEVEN

THE FUTURE

"Today is yesterday's pupil."

11.1 LOW-LEVEL MULTIPLEXING

An instrument per signal is costly. In many applications it is practical to use a multiplexer to process many low-level signals. This allows one gain-programmable instrument to do the work of many instruments. The multiplexer approach limits signal conditioning, but it does reduce cost.

A stream of multiplexed data can be digitized for immediate processing or storage. The data can also be left in analog form for observation or recording. The advantages of digital processing need no further discussion. It obviously can be used to supplant some of the conditioning provided by a dedicated instrument.

11.2 LOW-LEVEL FILTERING

Multiplexed signals without input filtering are subject to foldover or aliasing errors. RC filters are not very elegant and can be replaced by low-level active filters. The future will provide these filters for servicing signals prior to sampling.

11.3 COMPUTER-OPERATED INSTRUMENTS

The technology is available today to operate instruments by computer. Gains, offset, filtering, calibration, balance and so on, can all be remotely controlled. The discreet component approach to this control makes these instruments expensive. In the future, technology will continue to provide smaller components and increase the complexity in logic functions. These changes will reduce the cost of computer control. Eventually, the economic curves will cross and computer-operated instruments will be in more general use.

11.4 HIGH-LEVEL TRANSDUCERS

Efforts to build high-level transducers will continue. In some instances transducers are now available that can be used directly without further conditioning. The basic problems of common mode, excitation, and so on, still limits the application of these devices. Even high-level transducers need filters if they are to be sampled.

11.5 THE DIGITAL TRANSDUCER

After reading this book it is not surprising why people have a strong desire to avoid all analog processes. A data-gathering system that is fully digital would seem to be a good idea.

Transducers that store their output as digital data can be built. These data can then be interrogated over a bus system for further processing. There is merit to this approach, provided that the digital signal was properly derived. This approach squeezes the entire analog problem into the hands of the transducer manufacturer. The manufacturer must do it all.

11.6 DATA DECIMATION

The storage of all data taken during a test is not practical even though memory costs are constantly dropping. When all the data are stored, it costs money to extract pertinent information at a later time. Of course, if no analysis criterion has been established, the data must be kept until a decision is reached as to what to do with the data.

Data decimation, when it can be applied, is a powerful tool in reducing costs. In many situations simple algorithms could allow most of the data to be discarded.

As systems get more complex and as the number of data channels increases, a more formal approach to data decimation will be forthcoming.

11.7 THE ENGINEER AND THE TECHNICIAN

The world of instrumentation continues to grow. Engineers will be needed in increasing numbers to design systems and analyze results. Technicians will be the key in operating and servicing these systems. The engineering task must therefore include this problem of serviceability or else the technology will fall apart because of its own weight.

As the years go by we all get wiser. Today's designs are improvements over yesterday's. The question is often asked, "How did it ever work before?" Well it did! Perhaps all the shouting about how to do it better is not too relevant. This is obviously not true, however, because we thrive on seeing and using progress. It takes constant effort to keep the arrow pointed in the right direction.

Good education is always worth its cost. Engineers and technicians will continue to need this education to keep pace. It is hoped that this book will help provide some of this education.

INDEX